Intelligent Machining

Intelligent Machining

Modeling and Optimization of the Machining Processes and Systems

Edited by
Tugrul Özel
J. Paulo Davim

ISTE

WILEY

First published in Great Britain and the United States in 2009 by ISTE Ltd and John Wiley & Sons, Inc.

ISTE Ltd
27-37 St George's Road
London SW19 4EU
UK

www.iste.co.uk

John Wiley & Sons, Inc.
111 River Street
Hoboken, NJ 07030
USA

www.wiley.com

Library of Congress Cataloging-in-Publication Data

Intelligent machining: modeling and optimization of the machining processes and systems / edited by Tugrul Özel, J. Paulo Davim.
p. cm.
Includes bibliographical references and index.
ISBN 978-1-84821-129-2
1. Machining. 2. Machining--Computer simulation. I. Özel, Tugrul. II. Davim, J. Paulo.
TJ1185.I525 2009
671.3'5--dc22

2009009799

British Library Cataloguing-in-Publication Data
A CIP record for this book is available from the British Library
ISBN: 978-1-84821-129-2

Printed and bound in Great Britain by CPI/Antony Rowe, Chippenham and Eastbourne.

Table of Contents

Preface

Machining as a reliable manufacturing process still offers unmatched capabilities in producing high quality 3D parts from metals, polymers, ceramics, wood and composites. Advances in computational modeling and optimization methods have enabled researchers to develop cost effective and high throughput modern machining processes.

This book aims to provide recent advances in intelligent machining for modern manufacturing engineering. It includes six chapters on basic fundamentals, modern machining processes, analytical and mechanistic modeling approaches, finite element modeling and systems-based modeling, recent optimization methods and case studies.

The book can be used as a text book for senior level undergraduate engineering (e.g. mechanical, manufacturing, industrial, etc.) courses or at the postgraduate level as a book about machining. This book can also serve as a useful reference for academics, manufacturing and materials researchers, manufacturing and mechanical engineers, professionals in machining and related industries.

The Editors acknowledge their gratitude to ISTE-Wiley for this opportunity and for their professional support. Finally, I would like to thank all the chapter authors for their availability for this work.

Tugrul Özel
Rutgers University, USA

J. Paulo Davim
University of Aveiro, Portugal

Chapter 1

Modern Machining Process

1.1. Introduction

As a result of the advances in machine tools and cutting tool technology in recent decades, many of the conventional machining processes such as turning, milling and drilling have become cost-effective, flexible and high-throughput manufacturing processes for producing high precision and high quality discrete metal parts for the aerospace, automotive, die and mould manufacturing industries. These processes include: (i) hard part machining of hardened steels (or hard turning) into their near complete shapes, (ii) high-speed end milling at high rotation speeds, or "high-speed milling", (iii) high-throughput drilling. For example, until recently, high-speed milling was applied to the machining of aluminum alloys for manufacturing complicated parts used in the aircraft industry. In the past decade, with the advance of machine tools and cutting tool technologies, high-speed milling has been used for machining tool steels (usually hardness >30 HRC) for making moulds and dies employed in the production of a wide range of automotive and electronic components, as well as plastic molding parts [SCH 95, DEW 97, FAL 99].

Chapter written by Luis Norberto LOPEZ DE LACALLE, Joaquim DE CIURANA and Tugrul ÖZEL

In order to remain competitive in a global market environment, manufacturers should enhance the quality of their products and reduce costs while meeting strict customer requirements. Thus, recent research in the machining community has been mainly focused on increasing efficiency by fully utilizing the resources. It has been shown that actual machining times are much shorter than the non-productive times spent on loading/unloading, transferring, etc. the parts. Therefore, if consecutive operations can be performed on a single machine, it would decrease the production time and eliminate accuracy related problems due to re-clamping.

Major advantages of high-speed machining are reported as: high material removal rates, the reduction in lead times, low cutting forces, less work piece distortion and increased precision of the part. However, problems related to the application of high-speed machining differ depending on the work material and desired product geometry. The common disadvantages of high-speed machining are claimed to be: excessive tool wear, the need for special and expensive machine tools with advanced spindles and numerical controllers, fixturing, balancing the tool holder and lastly but most importantly the need for advanced cutting tools.

1.2. High speed cutting

High-speed cutting (HSC) of aluminum alloys has been a successful application of machining for several decades [SCH 82]. In the meantime, the advent of new cutting tools has made possible the machining of hard steels (usually hardness >20 HRC). Previously, the work piece used to be machined near-net shape in the annealed condition, heat treated to desired hardness, then machined to final dimensions and tolerances by grinding. Using advanced cutting tools it is possible to turn and mill steels in their hardened state to the final shape, thereby eliminating further grinding operations and the problems associated with work piece distortion.

Machining of alloy steels in hardened conditions, especially at high cutting speeds, offers several advantages. Those advantages are: reduction of finishing operations, elimination of distortion if the part is finish-machined prior to heat treatment, achievement of high metal removal rates, lower machining costs and improved surface integrity [TÖN 86, KÖN 93]. HSC of hard steels, however, results in high temperatures and stresses at the work piece-tool interface. Consequently, cost-effective application of this technology requires a fundamental understanding of the relationships between process variables on the one hand and tool life and machined surface integrity on the other hand. Thus, it is necessary to understand how temperatures and stresses developed during HSC influence tool wear and premature tool failure (or chipping) as well as residual stresses on machined surfaces. Therefore, a large number of research studies have appeared in manufacturing research journals over the past decade on the topic of high-speed cutting. In parallel, the demand particularly from the automotive industry for accelerated production during rapid globalization in 1990s pushed high-speed machining to its best.

Experimental data show that when machining hardened steels, work piece material microstructure (not only the hardness) and thermal properties affect the chip flow. It is common to observe higher cutting forces with higher work piece hardness. However, it is also observed that different thermal properties of the tool material may result in lower cutting forces [NG 97, ELB 97]. Therefore, in order to understand the process better and improve the performance of cutting tools, the use of deformation theory and advanced numerical techniques is recommended [STE 97]. Numerical modeling efforts and computational machining have become very popular to understand and further improve advanced applications such as hard turning, high-speed milling and high-throughput drilling.

1.3. Hard turning

The advent of cutting tool and machine tool technologies has made it possible to circumvent the traditional machining route for hardened steels (Figure 1.1). Previously, the work piece was machined to a near-net shape in the annealed condition, heat treated to the desired hardness, then machined to the final dimensions and tolerances by grinding. Using polycrystalline cubic boron nitride (PcBN) cutting tools it is possible to turn and mill steels in their hardened state to the final shape, thereby eliminating further grinding operations and the problems associated with work piece distortion. Therefore hard part machining (HPM) is often defined as machining of a material with a hardness higher than 45 HRC [DEW 97].

Figure 1.1. *Hard turning: cylindrical turning of hardened steel parts*

Turning of hardened steels into finished parts by eliminating intermediate machining and reducing grinding processes has been a cost-effective method for manufacturing high quality automotive components [BYR 03].

From a productivity viewpoint, HPM is reported to offer substantially higher metal removal rates than grinding, leading to a reduction in machining times [BUD 80]. Moreover, set-up and tool changing times are reduced. In contrast, HPM demands a minimum DOC (depth-of-cut), whereas in grinding an infinitesimal DOC can be removed by sparking out. Batch size is another relevant aspect in that, with large batches typically comprising 1,000 parts and more, grinding is generally the most suitable process. With small part numbers, however, HPM seems to be the best option. An economic analysis comparing grinding with OD (outside diameter) turning, chamfering and fine boring of hardened steels using PcBN tools indicate that the tooling cost per work piece for single point cutting is slightly higher than for grinding. However, when considering other matching costs, including capital investment, floor space and electrical power consumption, the cost for single point cutting operations is lower.

HPM offers the opportunity of eliminating the grinding process, which is usually applied after turning in order to obtain a good surface finish. If this can be achieved, benefits such as reduced machining costs and increased material removal rates can be obtained. Recent advances in cutting tool technology and the introduction of high quality lathes into the market have made hard turning a viable technology. The ability of manufacturing complex geometries in one machine and still obtaining a comparable surface finish as in grinding has made hard turning a favorable choice. Besides, since no or a minimal quantity of cutting fluid is used in hard turning, it is also beneficial from an ecological point of view.

However, a general approach to HPM is unrealistic and each application needs to be evaluated individually. Factors to be considered include: work piece type and condition, component design requirements including batch size, the machinery available and its condition, together with tooling and labor costs, etc. Tönshoff *et al.* [TÖN 86] outlined these factors and discussed the advantages of HPM in comparison to grinding in a comprehensive paper on the requirements of tools and machines when cutting hard materials. König *et al.* [KÖN 93] also compared turning vs. grinding under the aspect of surface integrity and attainable accuracy. Although, hard turning is a more flexible, more environmentally friendly and higher throughput alternative to grinding, process reliability and surface quality is still considered behind grinding processes due to issues related to geometrically defined cutting tools [KLO 05]. Figure 1.2 illustrates a comparison of capability profiles for hard cutting (turning) and grinding processes. Whilst dimensional and surface quality is superior with grinding over hard turning, process economics as well as flexibility were found to be less favorable.

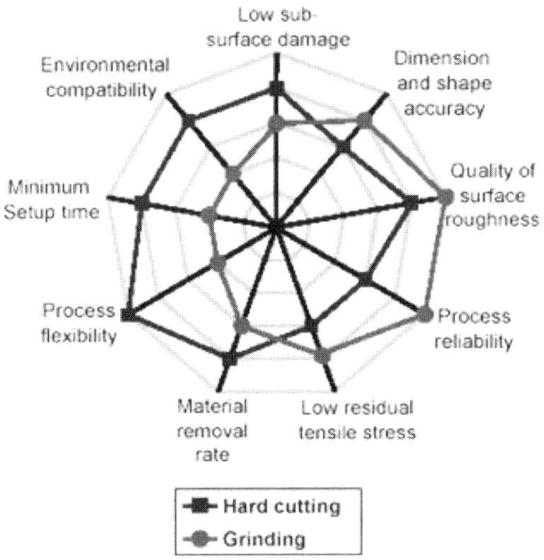

Figure 1.2. *Comparison of capability profiles for hard turning and grinding processes [KLO 05a]*

In hard turning, since hardened parts are machined, cutting tools are subjected to high stresses and temperatures and thus tool wear is an important issue. Different from the conventional turning, the thrust force, F_t, (Figure 1.3) component is the greatest in hard turning. This is due to the high hardness of the steel material and blunt shape of the cutting tool edges for added strength.

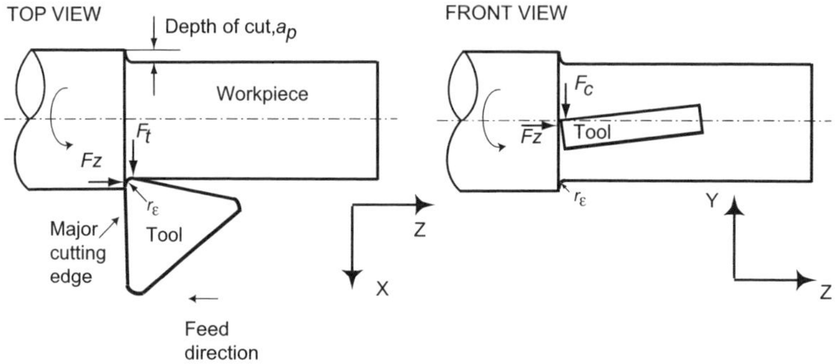

Figure 1.3. *Illustration of forces in 3D cylindrical turning*

In hard turning, PcBN cutting tools with various edge preparations (see Figure1.4) are preferred to protect the cutting edge, especially around the insert corner (nose) primarily, from chipping [BYR 03, ÖZE 03, ÖZE 05a, ÖZE 05b, KLO 05a, KLO 05b, MAT 99].

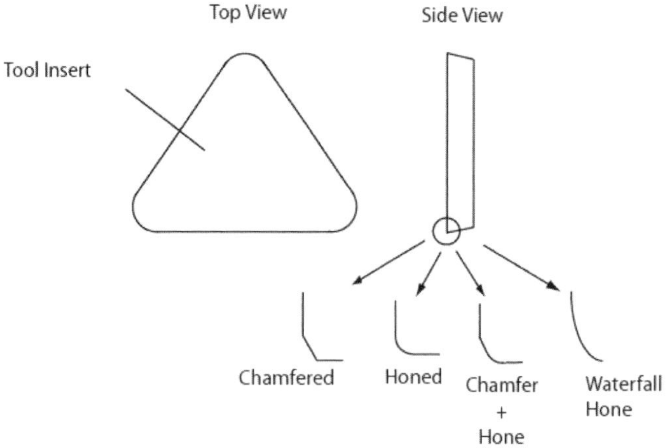

Figure 1.4. *Common edge preparations applied to PcBN cutting tools*

In order to resist tool wear under these harsh cutting conditions, cutting tool material should have high hardness, high toughness, chemical stability, high thermal conductivity and low thermal expansion coefficients. PcBN tools which possess most of the qualities given above are usually used in hard turning. Other than PcBN tools, cermet or ceramic tools are also used. PcBN tools are comprised of a polycrystalline layer of CBN crystals and a tungsten carbide substrate which are bonded together. PcBN inserts are produced through powdered metal processes. Wafers are cut into slices and brazed to a carbide insert, and then edge preparation is performed. The CBN content, grain size and distribution and composition of the binder material greatly influence the performance of the PcBN tool. Many studies have shown that PcBN tools with a lower CBN content perform better in hard turning applications [DAV 96, DAV 97, ÖZE 05a, ÖZE 05b, KLO 05a]. While low CBN content (40-70%) provides improved edge wear; high CBN content (70-95%) offers higher strength and toughness. In the meantime, thermal conductivity increases with the increase in CBN content of the PcBN tool, yielding higher tool temperatures, and affects the wear behavior of the CBN-binder matrix.

Hard turning is a technology-driven process and therefore depends on machine and workholding technology. In combination with PcBN cutting tools, the development of high precision lathes which have better spindle systems and higher structural stiffness, improved the outcome of the hard turning process significantly. As mentioned before, the elimination of the usage of cutting fluids is the bonus of hard turning. PcBN tools have made dry or near-dry cutting possible because they are durable enough to resist heat generated during cutting. The application of cutting fluids to the cutting zone makes tools vulnerable to thermal shocks which may cause sudden breakages. Therefore cutting fluids, which contain chemical elements harmful to human health, are usually not applied in hard turning.

The selection of cutting conditions in hard turning is very important and directly affects the performance of the hard turning process. For example, increasing cutting speed increases the temperature at the shear zone and tool-chip interface which softens the steel being cut due to the material's thermal softening behavior, thus reduced cutting forces are observed. Furthermore, low thermal conductivity of the PcBN tools assures that the heat generated is taken away and dissipated with the chips controlling thermal growth of the tool and the part. However, high cutting temperatures strongly influence the tool wear, work piece surface integrity and chip formation mechanism.

PcBN tools can withstand the extremely high temperatures but the binder material in the polycrystalline tool body fails after cutting temperature exceeds a threshold (usually around 1,200°C). Micro-geometry of the insert also has significant influence upon the surface integrity. Hence edge preparation must be carefully selected for a given application because it may generate subsurface damage and result in highly tensile residual stress on the surface of the machined work piece [KLO 05a, KLO 05b, DEN 05, ÖZE 07, ÖZE 08]. Heat generation during hard turning and heat distribution along the insert corner is also affected by micro-geometry due to a change in work material flow around the cutting edge. For example, a chamfered face provides excessive negative angle to the cutting action and results in high heat generation. PcBN tools rapidly wear out during hard turning at high cutting speeds mainly due to dissolving binder material at the high cutting temperatures attained. Flank wear occurs at the flank face of the minor cutting edge whereas crater wear mostly forms on the rake face near the major cutting edge [POU 01]. Often, flank wear is followed by crater wear and catastrophic tool failure (chipping). In order to improve the productivity and reliability of hard turning processes, optimum design of micro-geometry for the PcBN inserts must be accomplished.

1.3.1. *Tool micro-geometry*

It is without a doubt that the design of the cutting edge micro-geometry is becoming important to achieve high performance in machining with PcBN cutting tools. For high efficiency material removal with a corner (nose) radius (r_ε) insert, a constant uncut chip thickness to edge radius ratio ($\lambda=t_u/r_\beta$) should be maintained along the engagement area between the insert and the work piece. However, most edge preparations are applied uniformly all around the corner radius of the PcBN inserts. Uniform edge design hinders the potential of high performance cutting since it does not provide a constant uncut chip thickness to edge ratio (λ) along the insert corner in all engagement conditions. A uniform edge micro-geometry along the insert corner creates a very low uncut chip thickness to edge radius ratio at the minor cutting edge where uncut chip thickness becomes very small. This causes more ploughing than shearing at the minor cutting edge resulting in excessive stress and heat built-up, hence rapid tool wear. A variable edge micro-geometry along the insert corner as illustrated in Figure 1.5 and Figure 1.6 has the potential to reduce the stress and heat built up and enable hard turning at higher cutting speeds and feeds with less tool wear.

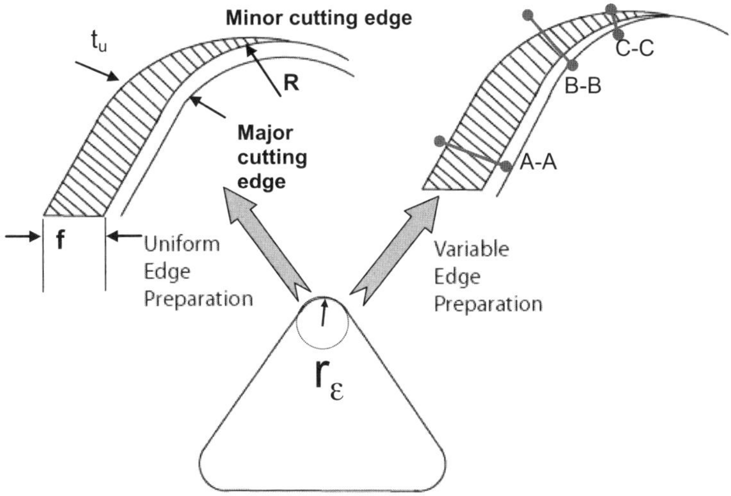

Figure 1.5. *Uniform vs. variable micro-geometry design [ÖZE 08]*

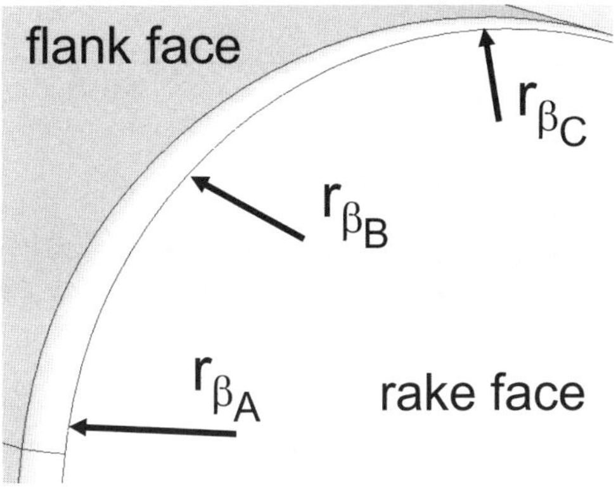

Figure 1.6. *CAD model of the variable hone edge design [ÖZE 08]*

While the mechanics and dynamics of machining hard metals using tools with uniform edge preparations have been analytically modeled [DAV 96, DAV 97, ELB 96, DAV 97, NG 97, KAR 08a, KAR 08b], finite element (FE) modeling studies have also provided essential information about the influence of edge preparation on the process variables such as chip formation, forces, temperatures and stresses [ÖZE 03, YEN 04, HUA 05a

CHE 06, ÖZE 07, COE 07]. Serrated chip formation is also revealed with increasing cutting speed and feed rate in most experimental and FE studies. Guo and Liu [GUO 02] proposed a 3D FE model for hard turning of AISI 52100 steel using PcBN tools. The model was used to predict the temperature distribution over the cutting edge, the residual stress distribution on the machined surface and cutting forces. Klocke and Kratz [KLO 05b] utilized 3D FE modeling to investigate chamfer edge design in wiper inserts, particularly identifying high temperature zone "hot-spots" on the tool face. Aurich and Bil [AUR 06] introduced 3D FE modeling for segmented chip formation. Karpat and Özel [KAR 07] presented experimental and simulation results for 3D FEA (finite element analysis) of hard turning using chamfered and honed inserts. Özel *et al.* [OZE 08] reported experimental and numerical investigations on the 3D hard turning process in the presence of variable design PcBN cutting tool inserts Experiments are used in the validation of numerical results. 3D computational modeling is utilized to predict chip formation, forces, stresses, temperatures and tool wear on uniform and variable edge design tools. Predicted temperature and stress distributions and tool wear contours reveal the advantages of variable edge micro-geometry design (see Figure 1.7).

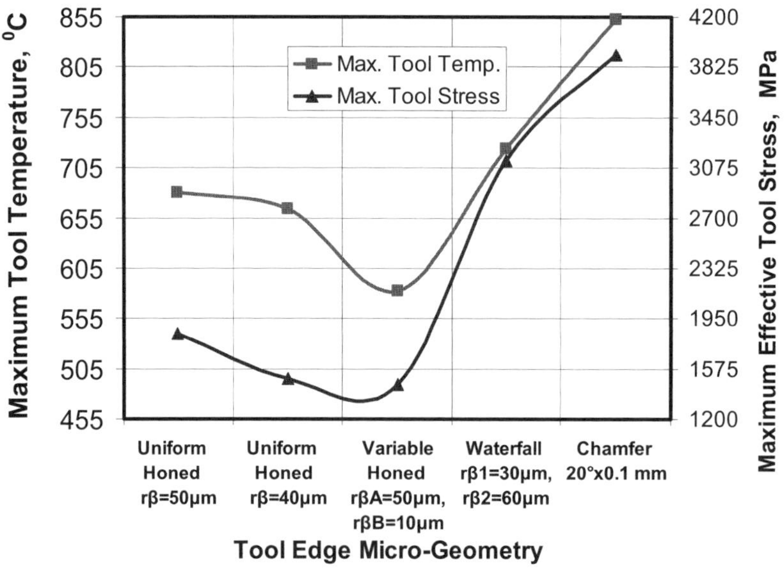

Figure 1.7. *Influence of micro-geometry on maximum tool temperatures and effective stresses obtained with FEA [ÖZE 08]*

1.3.2. *Surface quality and integrity*

Hard turning processes have a strong impact on the surface integrity. Surface integrity requirements can be categorized as geometrical surface integrity (surface roughness, dimensional accuracy) and physical surface integrity (residual stresses, microstructure changes) [MAT 99, KLO 05a, HUA 05a]. Surface roughness, which is maybe the most important customer requirement, is greatly affected by the selection of cutting conditions, tool edge geometry and tool wear. Residual stresses are the stresses which remain in the part after machining. The residual stress values in the work piece are mainly influenced by tool tip and tool-work piece interaction as well as the friction between the surfaces. Edge geometry of the cutting tools is an important factor on the machined surface integrity. Worn tools cause undesirable tensile residual stresses and increasing tool wear values push those residual stresses into deeper regions in the sub-layer of the work piece. Plastic deformation and thermal effects are crucial factors and should be considered in residual stress modeling. Residual stresses affect the fatigue behavior of the part under actual working conditions. Other than residual stresses, the formation of white layers, which appear as a white layer at the surface of the material under metallographic examination, characterizes the surfaces produced by hard cutting process. They can be caused either due to severe plastic deformation that causes rapid grain refinement or due to rapid heating to austenite temperatures and fast cooling (quenching). White layer formation can also be seen in grinding applications.

1.3.3. *Tool wear and failure*

Tool wear is an important problem for hard turning because it not only affects the economic advantage of hard turning, it also affects the surface quality of the work piece. The friction mechanism and heat generated between tool, chip and work piece determine the tool wear. The cost of PcBN tools are high compared to conventional cutting tools and the downtime during tool change has an additional cost which may hinder the economic advantage of the process. In order for hard turning to replace grinding, the relationships between surface quality, dimensional accuracy and tool wear should be very well analyzed. The tool performance is mainly influenced by factors such as: work piece material, cutting conditions (cutting speed, feed rate and depth of cut), tool material composition, tool geometry (rake angle) and tool edge preparation (i.e. micro-geometry of the tool tip). Increasing tool wear yields unsatisfactory surface finish and increased power consumption.

Wear is basically removal of material from the body of the tool and is a result of process parameters existing at the tool-work piece interface due to relative motion and cutting conditions. There are various mechanisms and modes of tool wear that have been observed in metal cutting. The investigation of machining wear due to cutting speed and temperature shows that tool wear is due to a complex combination of wear modes and is highly sensitive to cutting conditions (see Figure 1.8).

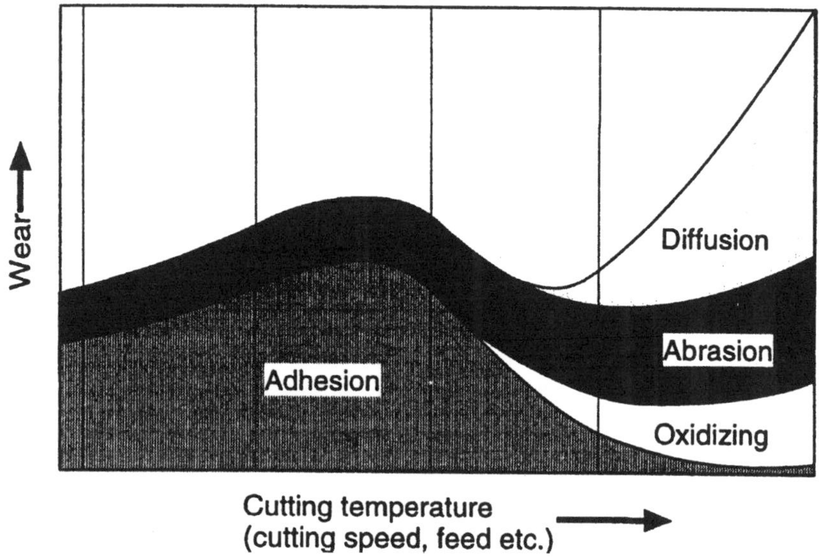

Figure 1.8. *Schematic diagram of the tool wear mechanisms appearing at different cutting temperatures corresponding to cutting speed and feed [KÖN 84]*

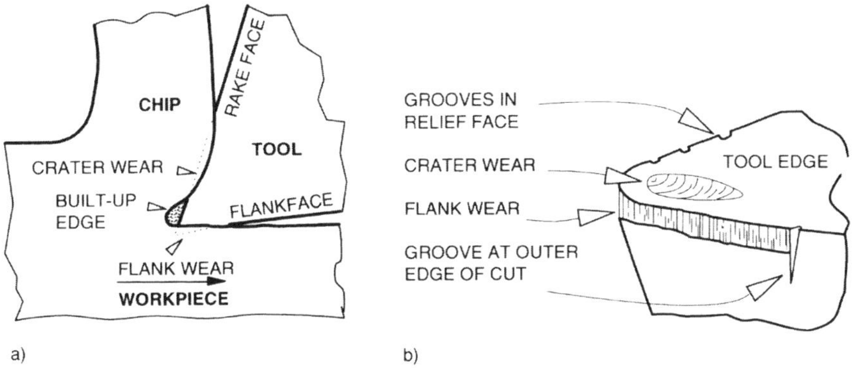

Figure 1.9. *Typical locations of location dependent wear on the tool [HOL 94]*

Flank wear at the front edge of the tool flank face and crater wear at the tool rake face (see Figure 1.9a) are the most typical modes of tool wear in hard turning. Flank wear is mainly a result of abrasion of the tool by hard particles (but there may also be adhesive effects); it is the dominating wear mode at low cutting speeds. Crater wear is the formation of a groove or a crater on the tool rake face where the chip rubs the tool surface (see Figure 1.9b). Different wear mechanisms, predominantly present in metal cutting, can be classified as sliding wear mechanisms and non-sliding wear mechanisms.

1.3.3.1. *Sliding tool wear mechanisms*

Sliding wear mechanisms include abrasive wear, adhesive wear and delamination wear. Abrasive wear is caused by the contact of a tool body with a particle, where the particle slides over the body and the movement causes material loss. Adhesive wear occurs between two surfaces rubbing against each other as a result of formation and breakage of the interfacial bonds. Delamination wear is another mechanism of wear when chips are sliding against tool surfaces at low speeds which do not generate very high contact temperatures. In this process, large wear particles are removed in the form of layers by the process of plastic deformation of the surface layer [SUH 73].

1.3.3.2. *Non-sliding tool wear mechanisms*

Non-sliding wear mechanisms include solution wear, diffusion wear, electrochemical wear and oxidation wear and become dominant with the chemical instability where cutting temperatures are very high: up to 1,600°C as in high-speed machining [SCH 95]. Solution and diffusion wear occurs when work piece and tool material react with each other to form a solution. The rate of the solution is dependent on cutting temperature at the interface and increases with the increasing temperature [SUH 86].

The wear due to diffusion is more dependent on the chemical properties of the material rather than its mechanical strength or hardness. Electrochemical wear is caused by a thermoelectric electromotive force (emf) generated at the work-tool junction during the cutting process. The emf causes electric currents to circulate and results in the passage of ions from the tool to the work piece [MOO 75]. Oxidation wear is a result of the chemical bonding of tool material atoms with oxygen. Oxidation in metal cutting occurs in the area where the chips are generated at elevated temperatures, usually above 700°C [KöN 84]. This type of wear is caused due to a chemical reaction between CBN's chemical composition and work piece materials. As an example, since it tends to accelerate chemical wear, ductile iron is not suitable for PcBN machining.

1.3.3.3. *Premature tool failure*

The sudden and premature tool failure caused by chipping of the tool cutting edge is one of the most important aspects of metal cutting and needs further investigation. In production with transfer lines, cutting tools should have a longer life than the predetermined period of time to facilitate tool changes at the end of the shifts. Premature tool failure not only damages the work piece, but also causes production downtime. In the past, chipping of the cutting edge was attributed to the impact loading resulting from the sudden engagement of the tool into the work piece material. Thermal fatigue wear is usually seen in interrupted machining where the cycle of heating and cooling stages results in thermal cracks and therefore tool breakage.

It must be noted that some of these wear mechanisms may become more dominant under different cutting conditions. Depending on the area where tool wear occurs, three types of tool wear pattern can be identified: flank wear, crater wear and chipping. Flank wear occurs between the tool and the work piece due to a rubbing action. In real life applications, the maximum permissible value of flank wear is usually used as a decision factor on tool life since it is easy to measure flank wear. It also has the biggest influence on the surface quality of the finished component. Crater wear occurs on the rake face of the tool because of chemical interaction between tool material and work piece. Chipping results from small broken pieces of the tool and is reduced by increasing the toughness of the tool. The tool wear patterns are shown in Figure 1.10.

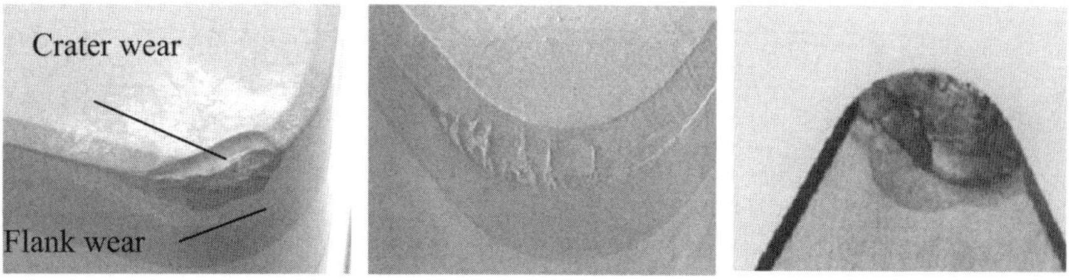

(a) Flank and Crater Wear (b) Thermal Fatigue Wear (c) Chipping

Figure 1.10. *Tool wear patterns in hard turning*

1.4. High-speed milling

The concept of high-speed machining and consequently the term *high-speed milling* (HSM) has changed in recent years, and thus a certain amount of historical background is required for a better understanding of the current meaning.

1.4.1. *Historical remarks*

The first reference to the possibility of machining at high cutting speed is to be found in Carl Salomon's patent [SAL 31], dating back to the late 1920s (German patent #523594). The essence of the patent was that tool wear and temperature increase along with cutting speed to a maximum value, the so-called *critical speed*; over this speed, temperature and wear may decrease along with speed, as Figure 1.11 shows. Therefore, machining carried out in excess of this value would obtain optimum tool life and productivity.

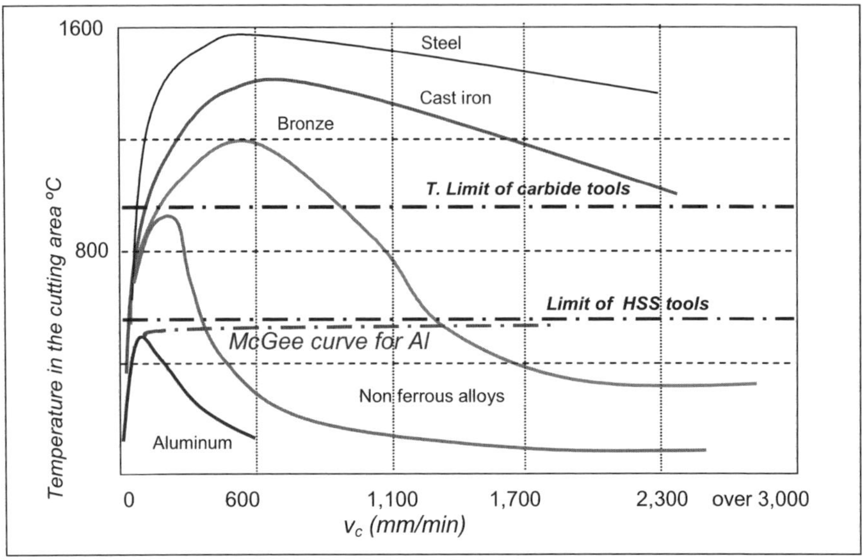

Figure 1.11. *Salomon's curves, cutting speed vs. tool/chip interface temperature*

Figure 1.11 also shows the McGee proposal [MCG 78]. For the reader's reference, in this figure the dashed lines show the maximum temperatures that current carbide and *high-speed steel* (HSS) tools can withstand.

All Salomon's work disappeared in 1945 during the Second World War, and so there is a wide gap in our knowledge of the experiments carried out during that period, although the patent does make a number of references to milling experiments at speeds in excess of 15,000 m/min.

New ballistic experiments were conducted in the 1960s and 1970s, shooting projectiles against material sheets in order to simulate the material deformation at high cutting speeds [KRO 61].

In 1979 a four-year project with a generous budget, known as the Advanced Machining Research Program (AMRP), commenced in the USA. The program [FLO 83][FLO 84] sought to reduce missile manufacturing costs and times by acquiring a scientific database on machining processes. The materials studied were steels, tool steels, lead, aluminum alloys, titanium alloys and others. A large number of experiments were performed, but we can quote certain conclusions [LEE 81] that are still valid today:

– cutting force decreases as cutting speed increases, to a minimum value at a cutting speed characteristic for each machined material. This fact is in agreement with Salomon's original idea;

– tool/chip interface temperature always increases along with speed, approaching the work material's melting point, contrary to the theories drawn up by Salomon. McGee [MCG 78] established the material's melting temperature as a threshold (see Figure 1.11);

– tool life always decreases as cutting speed increases, with no observation of the effects mentioned in Salomon's patent.

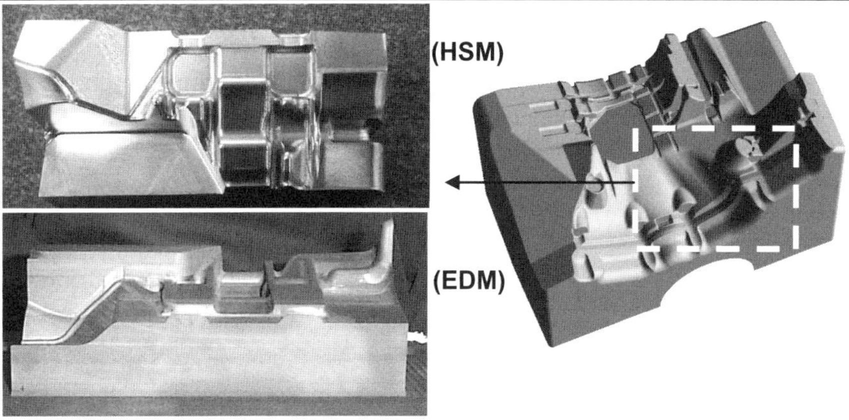

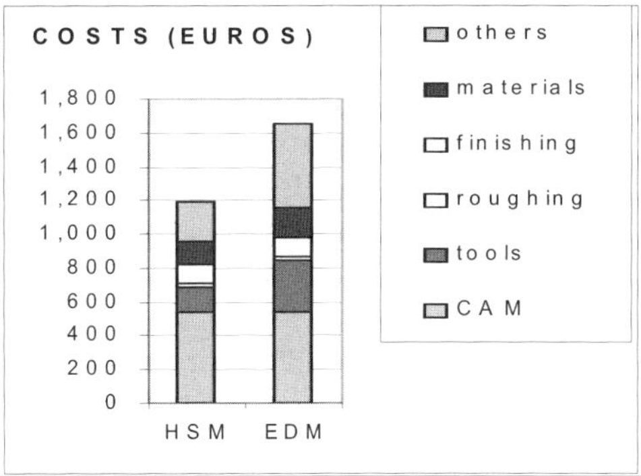

Time	*HSM*	*EDM*	
CAM	400	150	Hours
Roughing	802	802	Minutes
Finishing	1,174	2,280	Minutes
Sep up	30	702	Minutes
Others	120	150	Minutes
Electrode		1,420	Minutes
Total	26,126	14,354	Minutes

Figure 1.12. *A portion of an aluminum injection mould machined with HSM and with EDM. Top left, images of both pieces. Bottom, costs and process times for both technologies (this work was a collaboration between the University of the Basque Country and the University of Mondragon)*

Development was carried out in Germany in the 1980s on prototypes of high-speed milling machines [SCH 98]. The results obtained from German projects were as follows:

a) cutting speeds were increased tenfold;

b) new cutting tool grades were developed;

c) higher speed and acceleration were achieved on machining centers; and

d) skilled technicians were trained. Similar projects were carried out in the 1990s in France, Japan and the Basque Country (Spain) [LOP 02b].

In the course of the 1990s, HSM came into direct competition with *sinking electrical discharge machining* (SEDM) to manufacture forging dies and moulds. Several technological comparison studies [LOP 02a] on the technological benefits/drawbacks of one technique, with respect to the others, were performed. One example, shown in Figure 1.12, is that of a part of an aluminum injection mould chosen in view of its complex geometry. In this mould, roughing of the initial blank (250 HBN) was executed on a conventional machine. Finishing of two different rough parts was carried out on a high-speed milling machine and on an electrodischarge machine respectively, in each case after heat treatment on the steel (> 44 HRC). The required final roughness was 1.6 µm. Both time and cost balances (see Figure 1.12) were better for high-speed milling; time was reduced by 110%, and cost by 50%.

Today the techno-economical boundary between both technologies is well defined, thus, HSM is applied in all cases, except when extremely deep and narrow features have to be produced with milling. HSM's greater flexibility and shorter production times have now edged out the use of electrical discharge machining for the most complex surface parts.

1.4.2. *Current meaning of HSM*

After our brief historical narration, we will attempt to define the current meaning of HSM. HSM is a process whereby chips are considerably small since uncut chip geometry formed with the feed per tooth (*fz*), axial depth (a_p) and radial width (a_e) of the cut are small; however, cutting speed (v_c) is a lot higher (usually 10-20 times) than the values generally used in the 1980s and 1990s. In other words, each tooth of the cutter removes a small amount of material (the new chip), but does so extremely rapidly and hence efficiently.

With respect to the intrinsic nature of the material removal process, there are two possible cases of high-speed machining:

Case A: in the milling of light alloys (aluminum, magnesium and copper base alloys) current cutting tools allow cutting speed (v_c) above 2,500 m/min to be applied. At these high-speeds the physics of the material removal mechanism varies considerably with respect to the conventional speed in two ways:

– the strain rate at the chip primary shearing zone is greater than 10^5 s^{-1}; this is around 10^3 s^{-1} in conventional processes. However, work material becomes softer and easier to remove due to the higher cutting temperatures;

– most of the heat derived from the intense shearing deformation energy is evacuated with the chips removed and heat is not diffused towards the work piece, protecting the machined work piece from thermo-mechanical surface alterations.

Case B: for steels, cast iron, difficult-to-cut alloys, titanium alloys. Although the cutting speeds applicable are now much higher than those once considered conventional, there is hardly any variation in the intrinsic mechanisms (viscoplasticity, thermal balance, etc.) associated with chip formation. The process is faster but conventional in the intrinsic meaning.

The two definitions above, therefore, make it rather difficult to explain the concept of high-speed machining in exclusive consideration of the machining process itself. It does, however, make it somewhat easier to draw a comparison with the machining state of the art in the 1980s: we will thus use *high-speed machining* as the term for the process that applies a high cutting speed, i.e. cutting speed is appreciably higher than in the 1980s (over 10 times higher). Figure 1.13 shows the cutting speeds for each family of materials. As a general rule, a lower material machinability means lower cutting speeds. At the present time, as a matter of fact, increasing the cutting speed on titanium alloys from 100-300 m/min represents a greater technological leap than increasing speed from 2,000-4,000 m/min on a wrought aluminum alloy, for example.

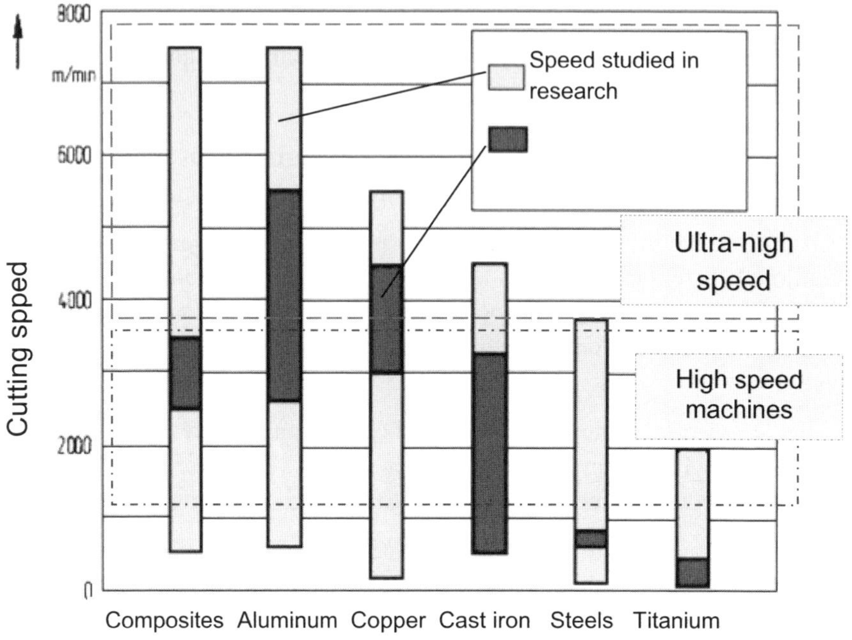

Figure 1.13. *High cutting speeds for different groups of materials. The darker color middle bands show the HSM cutting speeds commonly applied in industry*

As Figure 1.13 shows, there is a zone (the lower box) where the machine to be used is what is now called a *high-speed machine*, widely available from most machine tool manufacturers. Another zone (upper box) requires *ultra-high-speed machines*, which are still at the prototype stage and not yet available on the market.

HSM is part of high-speed machining. Milling is the most complex of the removal processes using a defined-edge tool (turning, milling, drilling, reaming, tapping, etc.) due to the following aspects:

– It is an *interrupted cutting* process. The tool's teeth suffer impact each time they enter the work piece. On the other hand, vibrations may be caused by the periodic variation of the cutting force.

– Due to the diameter of the mills (0.1-80 mm), which may be considered as small in relation to the size of the work pieces, and the high cutting speeds that can be used, milling machines must include main spindles that can reach 25,000 rpm with enough torque to remove material from the work piece. The high-speed spindle is a complex and expensive machine element. In turning, however, the work piece diameter is greater and higher cutting speeds are reached at relatively low headstock rotation speeds.

– The cutting tool must continuously follow a 3D tool path (*3-axis milling*) or be positioned with a certain orientation in the space with respect to the machined surface (*5-axis milling*). The milling machine has to interpolate several degrees of freedom to a high feed rate and therefore a powerful numerical control (NC) is required.

Another related concept commonly found in tool and machine catalogs and technical symposia is high performance cutting (HPC) – the main difference in relation to the classical approach of HSM is the high value used for the depths of cut a_e and a_p; several millimeters in the case of the former and only a few hundredths of a millimeter for the latter. This milling type can only be applied to low-machinability alloys such as wrought aluminum or light alloys (see section 1.4.3).

Fierce competition on the machine tools market, marketing strategies of companies and, in a general sense, the large number of technical references available nowadays (magazines, journals, Internet, machine and tool catalogs, books, etc.) have given rise to the usage of terms whose meaning is not so delimited. It is no longer as easy to make distinctions between what is high-speed milling and what is not. It might be simpler to distinguish between machines that are not high-speed: those with spindles producing less than 6,000 rpm and those that can generate over 12,000 rpm. In general, however, all CNC machines have increased their capacity in terms of accuracy and *material removal rate* (MRR) with respect to the machines of 15 years ago. The reader will perhaps permit us an example from another extremely competitive sector: nowadays, what cars are not "high-speed" or "high-performance cars"?

1.4.3. *High-speed milling of aluminum alloys*

With respect to aluminum alloys we may distinguish two different cases, depending on the type of alloys and component to be manufactured: wrought aluminum alloys and cast aluminum alloys.

1.4.3.1 *Aluminum alloys for airframe components*

Wrought aluminum alloys [WIL 96] [WAT 03] are widely used in the airframes of commercial aircraft. The most popular alloy type is the AAA7000 series (nomenclature of the American Aluminum Association) which is alloyed with zinc, magnesium and copper. The 7475, 7050 and 7075 are the most frequently used in this series with different heat treatments applied to them in order to ensure high ultimate strength. The AAA2000 series, alloyed with copper, is also a popular choice.

Wrought alloys are high-machinability materials with low *specific cutting force values* (700-900 N/mm^2). Therefore, the cutting speed can lie between 1,000-2,000 m/min and can be even higher if large-diameter milling discs are used. Within this speed range the material removal process gradually emerges as somewhat different to the conventional chip removal process, approaching the real meaning of high-speed machining explained in the previous section (Case A).

Structural parts such as wing ribs, stringers and longerons, spars, heavy frames, bulkheads and skins, which were traditionally assembled (using rivets) from simpler parts, are manufactured nowadays [AER 00] [AER 03] starting from a raw block machining up to their final shape (see Figure 1.14). Here it is necessary to remove up to 95% of the material creating airframe parts that are known as *monolithic parts*. These parts have enhanced homogenity with a higher strength-to-weight ratio. The *buy-to-fly* ratio (ratio of the weight of material purchased to the weight of the finished part actually added to the aircraft) of some components can be as high as 30:1, and therefore the chip removal rate must be as high as possible to make the milling of these parts economically feasible.

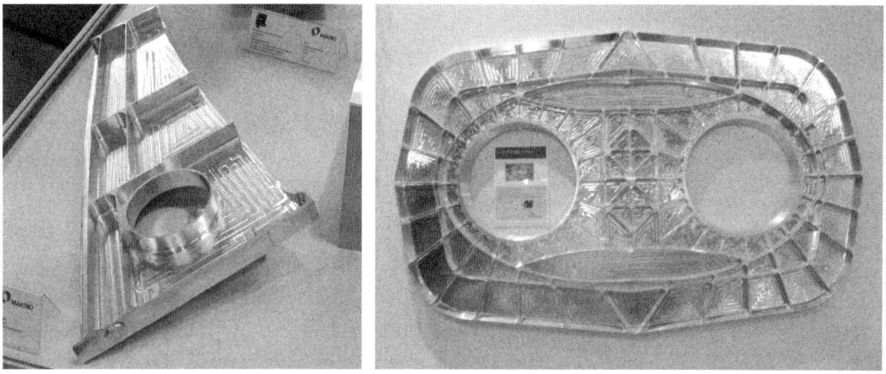

(a) (b)

Figure 1.14. *Airframe components made from a monolithic raw block: a) a rib and b) a bulkhead*

HSM also provides a better finishing quality due to the reduction of cutting forces at high cutting speeds [ISH 98] [SMT 98]. Pocketing at high removal rate conditions (i.e., high axial and radial depths of cut, high feed per tooth and high spindle rotation speed) is the typical process [AER 03]; the milling type is the HPC mentioned above. Spindle speed may be the maximum permitted by current industrial spindle technology; modern electrospindles are based on spindles with built-in asynchronous motors, whose shaft is supported by *hybrid bearings* (steel races with ceramic ball bearings) – the usual maximum speed is between 18,000 and 25,000 rpm and power is 30-60 kW.

With respect to cutting tools for the milling of aluminum alloys, these are the different choices:

– high-speed steels (HSS), the cheapest option, producing fairly good performance at cutting speeds of less than 800 m/min;

– sintered tungsten carbide without coating (WC), a more expensive solution than HSS, but with much higher performance. Today this is the "first option" in aluminum machining;

– hard metal with a TiAlN coating. Coatings improve mill performance in almost all cases. They are also extremely useful for machining in dry or near-to-dry conditions (see section 1.6).

Regarding tool concept and geometry, insert endmilling tools with diameters of 40-80 mm are used for hard roughing (see Figure 1.15a). In order to increase the tool/toolholder dynamics stiffness, integral shank-toolholders are available on the market (as shown in Figures 1.17a and 1.17b), and are capable of more than 40,000 rpm with a very good balance. Insert geometry is highly positive in all cases for effective chip removal.

In the case of a) small and medium-sized components, b) for geometrical details of large components or c) for finishing, integral bull-nose endmills are used (see Figure 1.15c) with diameters of 8-20 mm. The tool corner radius is usually 2.5 mm, so that it can reproduce the required corner radius at the wall-floor joints. Tool edges and tool shanks are relieved by grinding 6 or 8 mm up from the tool tip to prevent remachining at the top of high walls.

(a) (b) (c)

Figure 1.15. *a) Integral toolholder and shank with long inserts for pocketing, b) integral toolholder and shank for high rotation speed (30,000 rpm), c) a carbide bull-nose endmill with relieved shank*

Regarding the milling machines for this application, for large structural parts such as wings, fuselage skins, frames and stringers, machining centers with three long Cartesian axes (X, Y, Z) and with a 2-axis rotary headstock (axes A, C) are typical (see Figure 1.16). In this particular case the machine has two moving vertical spindles, so that any two components can be machined simultaneously or one component can be cut by two tools at the same time. In both cases there is an increase in the material removal rate.

Huge amounts of chips have to be removed from the working area during the milling of monolithic parts. To solve this problem, horizontal machining centers with a traveling table or traveling column are commonly used (see Figure 1.17). In these cases the headstock also provides two rotary axes. This configuration allows high chip volumes to be removed through a chip conveyor positioned between the column and the vertical front panel where the part is fixtured.

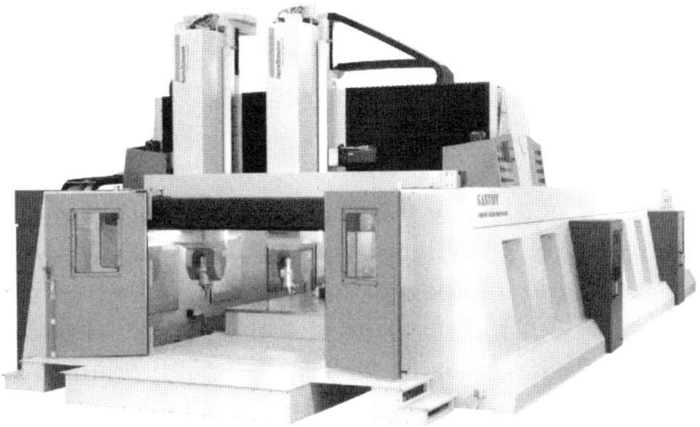

Figure 1.16. *Dual-spindle gantry mill by Handtmann®*

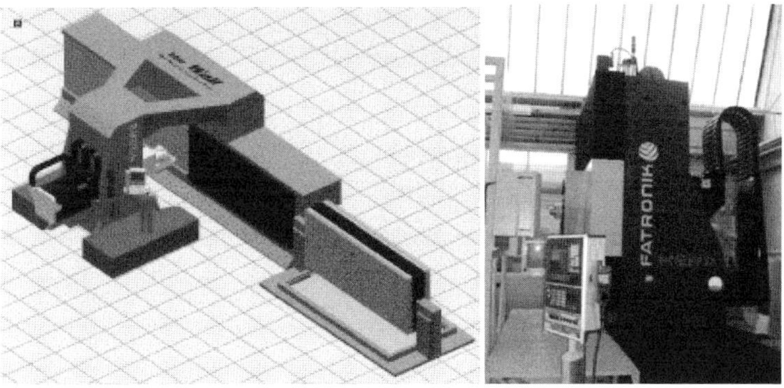

(a) (b)

Figure 1.17. *a) The wall machine concept by Mecof®, b) the Hera® machining center, based on parallel kinematics, by Fatronik®*

1.4.3.2. *Preventing vibrations induced by the machining process*

During the milling of monolithic parts, *forced* and *self-excited* vibrations may appear. These are a result of excitation by the cutting force of the "machine-spindle-tool" system (mainly during the roughing operations) or of the part (in the end milling of thin features) [HER 05] [BRA 05].

The periodic cutting force due to the tool teeth impact and cutting of each chip produces part and/or machine *forced vibrations* at the *tooth passing frequency*. The lower the cutting forces are, the lower the vibration effects become, and thus all due care must be exercised in relation to the maximum value of the cutting force. In the event of vibration of the machine-tool structure, high dynamic stiffness is recommended, and the spindle-tool system must also be carefully balanced.

On the other hand, *self-excited vibrations (chatter)* appear as the result of dynamic excitation produced by the chip thickness variation induced by the periodic irregular surface (waviness) generated by the preceding tooth. This variation adds a variable component to cutting forces, thereby exciting the structural modes of both the machine and the part. If the response to this excitation is not sufficiently damped it will be self-amplified, chatter will increase and machining will become unstable. Chatter brings about a very bad surface finish and can lead to a sudden breakage of the spindle hybrid bearings – it must be borne in mind that these are bearings with fragile ceramic balls running on steel races.

Chatter depends on rigidity, damping coefficient and natural frequencies of the machine tool, the cutting tool and the part. The so-called *stability lobes* diagrams can be calculated to prevent chatter. In these diagrams the borderline between stable and unstable cutting conditions is defined in terms of the axial depth of cut a_p and the spindle rotation speed in rpm (see Figure 1.18). The maximum removal rate conditions are obtained from the lobe analysis, as the "sweet point" marked in the figure.

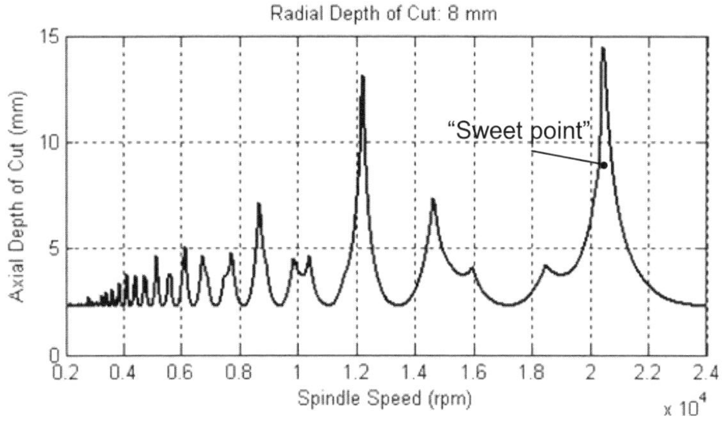

Figure 1.18. *Stability lobes present the rotational speed vs. axial depth of cut, in this case for a 50% tool diameter immersion. The zone under the borderline is the stable milling zone. A "sweet point" is an optimum point for very high MMR without chatter*

1.4.3.3. *Thin walls and thin floors*

In *thin wall* milling, deformation of the wall caused by the cutting forces on the free top section leads to a smaller radial depth of cut, and therefore a thicker final section, on the free top section; more than a tenth of a millimeter for a 1 mm thick wall has been observed [HER 05]. In *thin floors* the problem is similar at the center of vanes with no back support. For the milling of these high flexible features there are two effective solutions: firstly, using special workholdings and fixtures to make walls and floors stiffer, and secondly applying small depths of cut to produce a small chip section and therefore low cutting forces.

A useful machining procedure is the so-called *jump to jump.* This method uses alternative tool passes on both sides of the wall in order to constantly secure higher local rigidity at the cutting point. During every tool path, the section just below the tool contact point is the stiffest possible (see Figure 1.19a).

Moreover, the emergence of vibration during machining caused by local excitement of walls and floors has recently been covered in a number of research papers [BRA 05] [CAM 07]. The models used are similar to those applied in the event of vibration of the machine and/or the toolholder and/or the tool, albeit with an added problem; during machining of the thin floor or wall, part mass and stiffness vary, and thus so do the part's natural modes and frequencies, thereby making the *a priori* prediction of the stability lobes a more complex task.

(a) (b)

Figure 1.19. *a) Milling of a wall 0.8 mm thick, b) marks at the top of the thin wall caused by chatter vibration*

1.4.3.4. *High silicon-aluminum alloys*

Cast aluminum alloys, mainly used for car engine components (housings, blocks and pistons), are extremely abrasive because they contain silicon. Eutectic composition for the aluminum-silicon alloy is around 12% – in excess of this concentration, silicon is precipitated as primary crystals.

In engine components (see Figure 1.20) machining operations are simple because parts are cast to very precise shapes, with only material of 0.5-2 mm left as overstock. Therefore, only a) face milling with low axial depth of cut, b) drilling and reaming of holes and c) some square milling operations are required, with a low material volume to remove per part.

In HSM, large-diameter milling discs with inserts are the high productivity solution. Depending on the amount of silicon in the alloy, the tool material could be as follows:

– in alloys with less than 12% silicon, machining may be carried out with carbide tools or inserts coated with multilayer TiAlN coatings (see Figure 1.21). Cutting speeds can be between 600 and 1,000 m/min. PCD (polycrystalline diamond) is a useful but expensive option to increase material removal rate and tool life;

– in alloys with more than 12% silicon, only PCD tools may be used since carbide tools wear out over very short periods of time. The cutting speed with PCD is around 300-400 m/min.

Figure 1.20. *a) Typical aluminum-silicon injected part, b) hole-making operations on a cast aluminum component by Heller®*

Milling centers for these types of components are small or medium-sized (typical workspace is a cube X:1, Y:0.8, Z:0.8 m) 3-axis machines, generally with horizontal headstocks capable of more than 18,000 rpm and providing 15-40 kW. In horizontal machining centers, an additional B-axis on the machine table allows the part rotation to work on all part lateral faces.

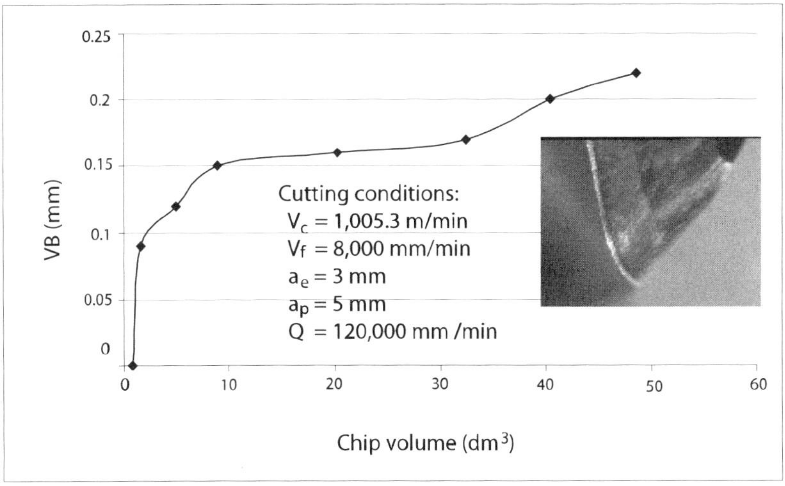

Figure 1.21. *HSM of a eutectic alloy (AlSi12CuNi1) with a Ø16 mm carbide tool coated with TiAlN. After 50 dm^3, flank tool wear V_B was 0.24 mm*

1.4.4. *High-speed milling of titanium alloys*

Alloy Ti6A14V is the reference when considering the machining of titanium. Depending on the heat treatment applied, two metallurgical states are commonly used in parts, annealed (σ_u: 900 MPa, 36 HRC) and aged (σ_u: 1170 MPa, 41HRC). The former is usually milled whereas the latter is finished and drilled.

The following considerations may be made [KOM 82] [BRI 97] in relation to the machining of titanium alloys:

– the poor thermal properties of titanium alloys lead to segmented and serrated-like chips;

– the high temperatures in the tool/chip contact area and the high chemical reactivity of titanium with almost all tool materials are the main reasons for the rapid tool wear. This commences at the rake face as *crater wear*, but it is in very close proximity to the tool edge, thereby accelerating the *flank wear*;

– the low elasticity modulus of titanium alloys causes part deformations and vibrations during the milling of thin parts.

The current use of micrograin carbide tools dates back to the 1950s, and now the K10 grade is used as reference. Cutting speed ranges from 80-300 m/min and moderate chip section parameters are applied (*fz* 0.05-0.15 mm/z, a_p 2-4 mm, a_e ¼ ∅). For example, tool life with respect to cutting speed is shown in Figure 1.22. Emulsion coolant must be used at all times. Studies were performed on cutting materials based on compounds of rare earth metals [KRA 93], new submicrograin or ultramicrograin grades, new coatings, and assisted-cutting methods by high pressure coolant or liquid nitrogen cryogenic. In *high-pressure coolant-assisted milling*, emulsion coolant is injected between the tool rake face and the chip generated, fragmenting the chips and reducing tool wear [KOV 95].

A titanium-orientated machining center does not require a high-speed spindle because the maximum speed of rotation is around 6,000 rpm. One good solution is a main spindle directly connected to the main motor by a flexible coupling (the *direct drive* concept, see the next section). Currently 5-axis machines with a cradle table, i.e., two rotary axes on the machine plate, are widely used for blade and impeller milling.

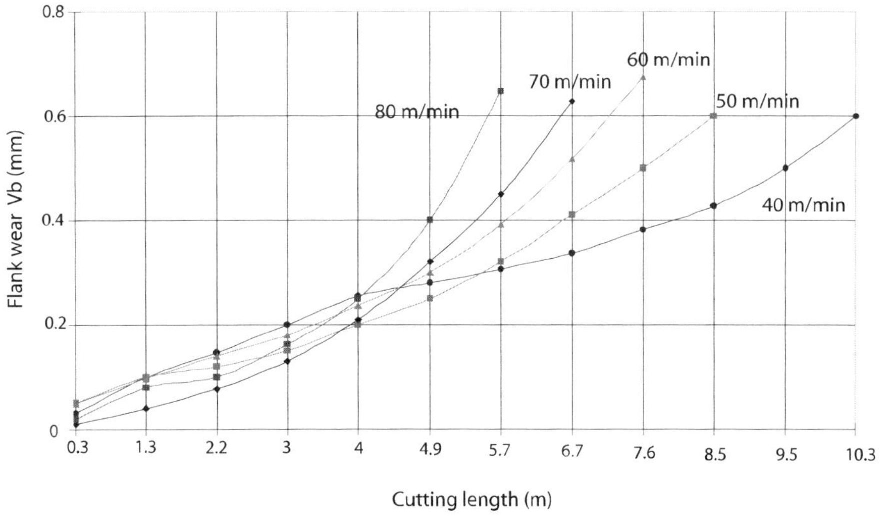

Figure 1.22. *Cutting length at different cutting speeds when milling Ti-6Al-4V annealed (∅20 mm endmill K10, z 3, f_z 0.15 mm, a_e 5 mm, a_p 3 mm)*

1.4.5. *High-speed milling of die/mould steels*

Forging dies, injection moulds for aluminum and plastic parts and stamping dies are manufactured from different materials (see Table 1.1), but all cases have two features in common. Firstly, the functional surface is a *sculptured* or *free-form* surface, and secondly material hardness is either high (30-40 HRC) or very high (40-60 HRC). In view of the former feature, the term *sculptured surface machining* (SSM) [LOP 08] is widely used to describe this technology.

Sector	Raw material	HRC	Roughness Rt	Tolerances	Fillet radius
Forging dies	56NiCrMoV7 (L6) X40CrMoV5/11 (H13)	42-60	5-10 μm	0.2 mm	More than 1.5 mm
Plastic injection Moulds	X37CrMoSiV5 1 (H11) X40CrMoV5 1 (H13) X30WCrV9 (H21)	48-50	1 μm (0.5 Ra), mirror-like finishing	0.01 mm in interfaces	Sharp edges
Aluminum injection Moulds	X40CrMoV5 1 (H13)	50-55	1-2 μm, mirror-like finishing	0.05 mm	Several cases
Stamping dies	GG25 (ASTM 48 Grade 40B) GGG70 (ASTM 100-70-03)	220-270 HBN	20 μm	0.1 mm in free forms	Small
Blades for aeroengines	Ti6Al4V	35	2-5 μm	0.05 mm	3 mm, but external

Table 1.1. *Requirements of different sectors involving complex surfaces*

For moulds and forging dies, SSM commonly involves the removal of a 0.2-0.5 mm allowance [LOP 02a] [SAL 05]. Due to the intricate shape details, ball endmills with diameter less than 20 mm, made from micrograin carbide and coated with AlTiN or AlCrN, are typical tools. The cutting speed ranges from 300-400 m/min.

Taking into account the fact that slopes commonly found in sculptured forms range from 0-90°, and that the effective cutting speed (i.e., the speed at point A in Figure 1.23) can be between 300 and 400 m/min with current tools, the spindle's speed must be over 15,000 rpm, thereby requiring a high-speed spindle for the machining center. Nowadays, the maximum rotation speed of industrial high-speed machines is around 20,000-25,000 rpm, with power output 14-20 Kw. On the other hand, for this speed of rotation, bearing in mind a recommended feed per tooth of around 0.07-0.1 mm/tooth, the machine's maximum linear feed is 10-15 m/min. These feed rate values can be obtained by typical linear ball screws connected to synchronous motors, or by linear motors. The control of interpolated curves at this feed is not a problem for current numerical controls [SIL 06].

In view of the thermophysical nature of the removal process, the HSM of tempered steels at 300-500 m/min is similar to lower values, and therefore it is not a true high-speed machining process. It is a “faster process” but of the same nature of conventional cutting speeds (Case B, see section 4.2). It does, however, require a high-speed machining center.

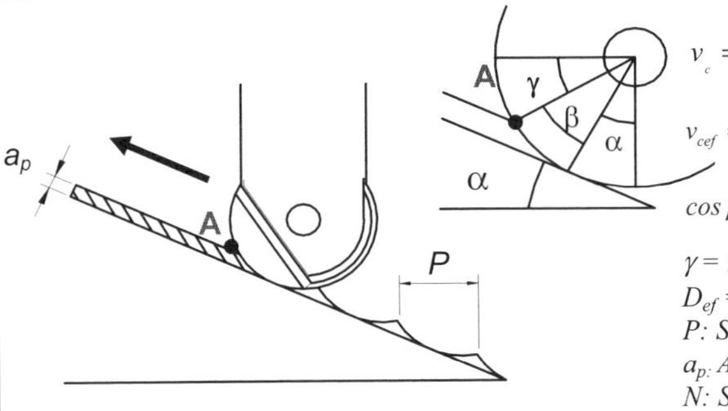

$$v_c = \frac{\pi \cdot D \cdot N}{1000}$$

$$v_{cef} = \frac{\pi \cdot D_{ef} \cdot N}{1000}$$

$$\cos\beta = \frac{R - a_p}{R}$$

$$\gamma = 90 - \alpha - \beta$$

$$D_{ef} = D \cdot \cos\gamma$$

P: Step or radial width of cut

a_p: Axial depth of cut

N: Spindle speed

Figure 1.23. *Milling of an inclined plane and formula relating spindle speed (N) to cutting speed (v_c)*

Machining centers for forging dies and moulds are usually 3-axis. Three solutions for the main motion (headstock) are possible, as shown in Figure 1.24:

– motor and spindle connected by a timing belt, the cheap option used in conventional machines with rotation speed lower than 6,000 rpm. Nowadays, this option is not used in high performance machines;

– main motion consisting of a motor directly connected to the spindle by a flexible coupling, the so-called *direct drive*. This configuration allows up to 12,000-14,000 rpm; its main advantage is its high reliability. This solution is also used in high precision machinery due to all the spindle thermal dilatation and can be absorbed by the flexible coupling;

– compact electrospindles: these usually provide 15-25 kW with maximum rotation speed 18,000-25,000 rpm. They use hybrid bearings metal races with ceramic balls – this is the main drawback, since even a small blow can damage the ball bearings.

Modern electrospindles can also be used for roughing tempered steel even at low rpm (< 1,000 rpm), since they provide maximum torque even at very low speeds of rotation.

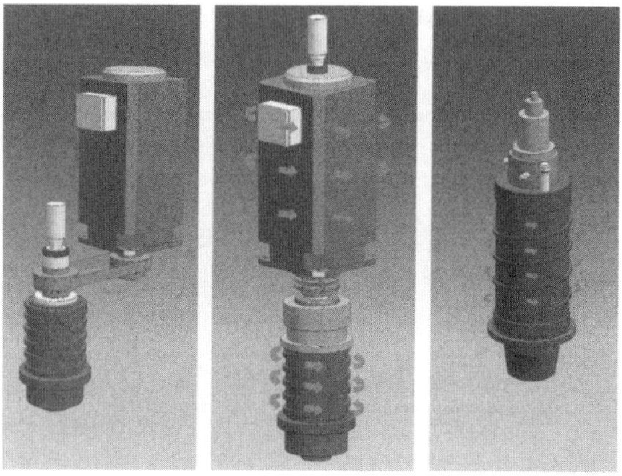

a) b) c)

Figure 1.24. *Three solutions for the machine tool's main spindle: a) the conventional, motor and belt, not recommended for high-speed operations, b) direct coupling, c) electrospindle*

Two applications of HSM on hardened steels are presented in Figures 1.25 and 1.26. In the former, a forging die, roughing and finishing were carried out directly from a tempered blank in one part fixturing, with a total machining time of 206 minutes. 30% of this time was performed by tools of diameter less than 2 mm, because the fillet radius was 1 mm in the deepest areas. The final roughness and precision were better than that usually required for forging dies.

As shown in Figure 1.26, a plastic injection mould made of 35 HRC steel was machined on a 5-axis HSM center. A bull-nose tool (toroidal tool) with corner radius 2 mm was used successfully. The part side flanks were machined using a *z-level* in a simultaneous 5-axis interpolation (usually known as *flank milling)*, finishing them in only 6 minutes; in 3-axis mode with a ball endmill, the same part took over 50 minutes.

Nomenclature: DIN 56NiCrMoV7 (AISI L6)
Hardness: 45 HRC
Initial blank dimension: 120x130x40 mm

a)

b)

Operation	N (rpm)	F (mm/min)	a_p (mm)	a_e (mm)	Tool
1. Rough	8,900	2,000	0.5	1	End mill 6mm
2. Rough	18,000	2,000	0.5	0.9	Ball endmill 6 mm
3. Rough	18,000	1,400	0.3	0.3	Ball endmill 3 mm
4. Bitangency roughing	18,000	1,400	0.2	0.3	Ball endmill 3 mm
5. Bitangency roughing	18,000	900	0.5	0.3	Ball endmill 2 mm
6. Semifinishing	18,000	1,400	0.2	0.3	Ball endmill 3 mm
7. Semifinishing	18,000	1,600	0.2	0.6	Ball endmill 4 mm
8. Bitangency Semifinishing	18,000	900	0.2	0.3	Ball endmill 2mm
9. Finishing	18,000	1,100	0.3	0.17	Ball endmill 2 mm
10. Bitangency Finishing	18,000	1,100	0.3	0.17	Ball endmill 2 mm
11. Finishing	18,000	1,400	0.3	0.17	Ball endmill 3 mm

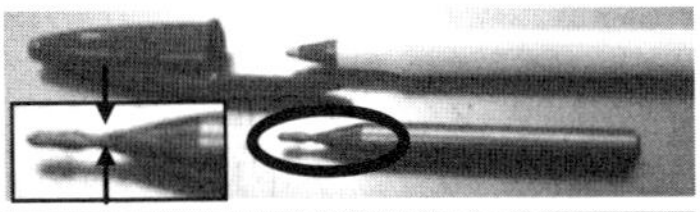

c)

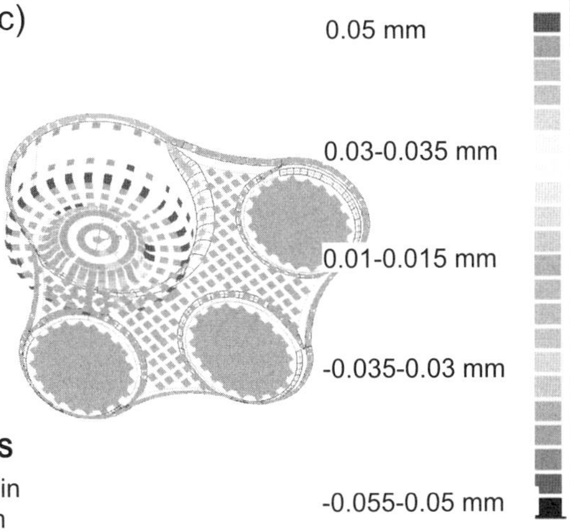

d) **RESULTS**

Roughing time: 124 min
Finishing time: 82 min
Roughness: 0.2 μm *Ra*

Figure 1.25. *High-speed milling of a die for the suspension of a car ball joint: a) constitutive material, b) machining operations and cutting parameters, c) details of the ∅2 mm ball endmill, d) results: final die, machining times, roughness and dimensional precision map (right)*

a) Mould steel Thyroplast© 2311 35HRC (AISI P20)

b)

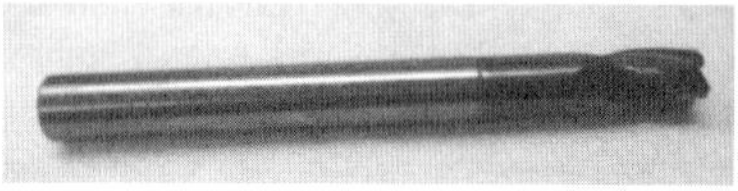

d)

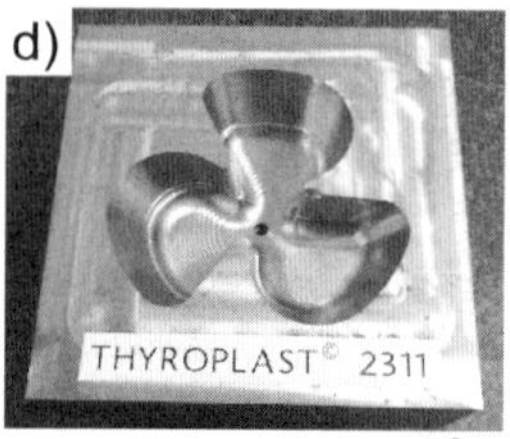

e)

c)

Tool	Operation	Vc (m/min)	fz	S (rpm)	Time (min)
Bull Nose ∅12 r2	Rough milling	234.4	0.2	6,200	43
Bull Nose ∅12 r2	Rough milling	234.4	0.12	6,200	11
Bull Nose ∅12 r2	Flank finishing	377	0.1	10,000	6
Bull Nose ∅12 r2	Floor finishing	377	0.1	10,000	8

Figure 1.26. *(a) Mould material, (b) bull-nose endmill ∅12, (c) machining operations, cutting parameters and machining times, (d) part geometry, e) 5-axis HSM milling machine by Ibarmia®*

1.4.6. *High-speed milling of superalloys*

Nickel superalloys are used for turbine vanes, disks, combustion chambers, exhaust valves, furnace equipment, pre-combustion chambers in diesel engines, rocket engine components and several components used at nuclear power stations. The most widely used alloy is the Inconel 718 (AMS 5596, UNS N07718, 52Ni-19Fe-18Cr-5(Cb+Ta)-3Mo-0.9Ti-0.5Al), with two metallurgical states, annealed (hardness over 28 HRC) and aged (over 40 HRC).

Nickel-based alloys are considered to be the lowest-machinability metal alloys for three main reasons [VIG 97]:

a) nickel-based alloys are ductile materials; this fact implies a high chip-tool rake face contact area. Tool edge chipping is likely to occur due to the consequent high cutting forces;

b) burrs frequently emerge on part edges being a risk for user and requiring a subsequent deburring operation;

c) these alloys have a *mechanical hardening* tendency caused by their austenitic matrix.

Carbide endmilling tools with TiAlN coating are widely used, applying moderate values for cutting speeds (60-100 m/min) and feeds (0.025-0.07 mm/tooth). Neutral or even negative rake angle tools are used. Figure 1.27 shows the results of tool life for different cutting speeds, for a 0.025 mm feed per tooth.

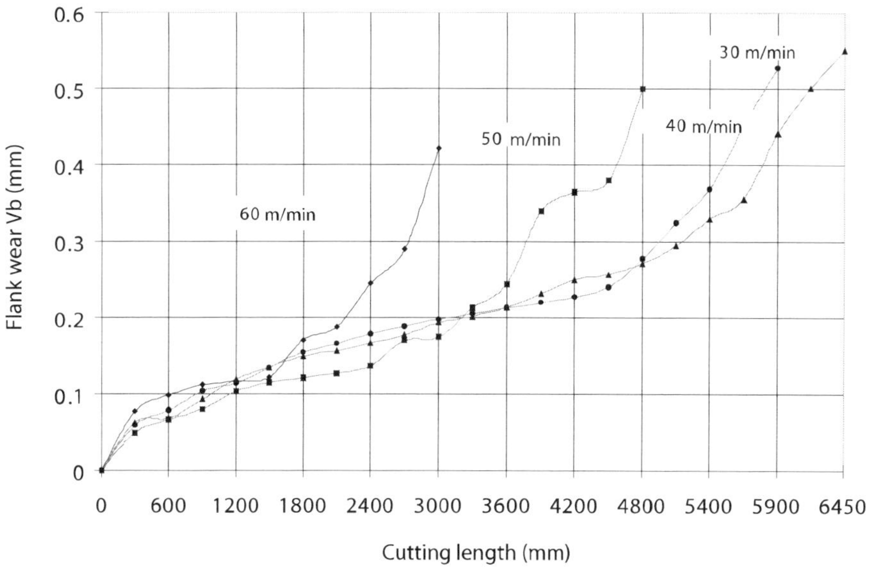

Figure 1.27. *Cutting length vs. tool flank wear in the machining of Inconel 718 (annealed), for different cutting speeds; K10 Endmill ∅20 mm, a_p and a_e were 2 mm, fz 0.025 mm*

In the machining of superalloys, *thermally enhanced machining* (TEM) [LOP 04] has been tested, but it is a long way from being accepted in industry due to material integrity considerations.

1.4.7. Machine tool technology

The stringent requirements in terms of time and precision for complex parts on one hand and the power of numerical controls on the other, along with the creative mindsets of designers, have introduced some new concepts for machine tool construction. Thus, a new view of a machine tool is to consider it as a servo-controlled spatial mechanism that guides and drives a cutting tool along a complex trajectory creating a new shape in the raw material; all current advances in mechatronics, mechanics, materials, control, sensors and communications have been added to machines.

Several aspects of current high-speed machines have already been mentioned in previous sections, such as the machine structure (see 1.4.3.1) or the spindle options (Figure 1.24 in section 1.4.5), but they need to be extended and other issues must also be addressed.

1.4.7.1. *Milling machine structure and materials*

Milling machines are 3, 4 or 5-axis spatial mechanisms [LOP 08]. The spatial configuration of axes may differ between models, but 3 basic types are common in HSM:

1. *Fixed bridge* structure, with moving table in X, moving slide in Y and vertical embedded ram in Z. This is very common in small and medium-sized centers. In large machines the X-axis is solved by a traveling beam on the bridge (Figure 1.28, 1st row), while the part remains motionless;

2. *Traveling column,* with all the Cartesian axes in series to move the tool. The main advantage is that the mass to move is always the same, and does not depend on part size. On the other hand, the long table allows milling to be carried out in two zones (see Figure 1.28, bottom row), rendering the work piece fixturing easy in one zone while the machine is acting on a part in the other.

3. The "box-in-box" structure is an extremely typical configuration of horizontal centers due to its considerable stiffness and mechanical and thermal stability (see Figure 1.29).

In addition, the resultant force introduced by the axis drives is recommendable to be applied at the center of gravity of the moving components [MOR 08]. Typical values of acceleration at the tool center point are between 0.8 and 2 g, and 1 g is an extremely common value for medium-size machines.

Machine stiffness is extremely important to retain the desired precision withstanding the high cutting forces, the inertial loads and the weight of the work piece. Typical stiffness values measured at the spindle nose are between 20 and 60 N/μm [LOP 05].

The accuracy of a machine tool is a key factor in the production of high-added value parts. Figures vary between models, and in addition there are several ways to measure this, with many determining factors [LOP 08]. The most recent reviews [BYR 03] give values of 4-8 μm for high quality machines, and even 1 μm for precision machine tools of conventional size. Accuracy in the case of micromilling machines is even higher.

Materials other than steel and cast iron are not conventional for structural components (beds and moving components). Granite and polymer concrete, however, are increasingly used. Polymer concrete (also known as *mineral casting*) is a combination of mineral fillers bonded using a resin system. Polymer castings are suited to precision machines due to their extremely low thermal diffusivity, while granite is suitable for very accurate machines such as high precision milling machines or grinders. Research projects now focus on the use of fiber-reinforced composites or bar structures [NEX 08], but only a handful of prototypes have been developed.

1.4.7.2. *Drive trains*

Today linear motors make longitudinal displacements possible without other transmissions, but ball screws are still a cheaper and more efficient solution.

In the better ball screw solution for HSM, the feed motor, usually a synchronous brushless motor, is connected to the screw by a flexible coupling, which is a good torque transmitter and heat insulator, and this eliminates the need for perfect motor alignment with respect to the screw. The connection of the motor to the screw by a timing belt is cheaper, but not appropriate for high accelerations. Screws are supported at both ends by angular ball or conical roller bearings, in double-back (DB) configuration in the case of the former, to increase the rigidity of the screw. To the same end, screws are usually pre-stressed, which also reduces the impact of thermal dilation. The machining component to slide is bolted onto the nut.

On the other hand, iron-core synchronous linear motors present very good dynamic behavior. In these motors, a coil slide acts as *primary* and a magnet plate as *secondary* on a linear basis. One of these is fixed to the moving component, while the other is fixed to the machine bed. The main advantage of linear motors is the absence of mechanical elements of transmission. Another important advantage is the absence of natural resonance frequencies in the drive. Linear motors are now improving some disadvantages such as heat evacuation and their prices are falling, thus their use in high-speed machines will gradually increase.

1.4.7.3. *5-axis milling machines*

5-axis machines allow the rotary tool, a cylinder with five degrees of freedom, to be moved in any orientation and position in the workspace, reaching a specific orientation with respect to each work piece surface.

Three configurations are common in 5-axis milling centers, shown in Figure 1.28; the kinematic chain will be defined starting from the work piece towards the tool tip, where *L* denotes a linear axis and *R* a rotation axis:

a) the LLLRR focuses on large parts, for example stamping dies or aeronautical fixtures;

b) RRLLL is used for small and medium-sized components and it is very common in moulds, impellers, biomedical components and high-added value parts machining; finally,

c) RLLLR is recommendable for tall work pieces, e.g. wind generator components, for example.

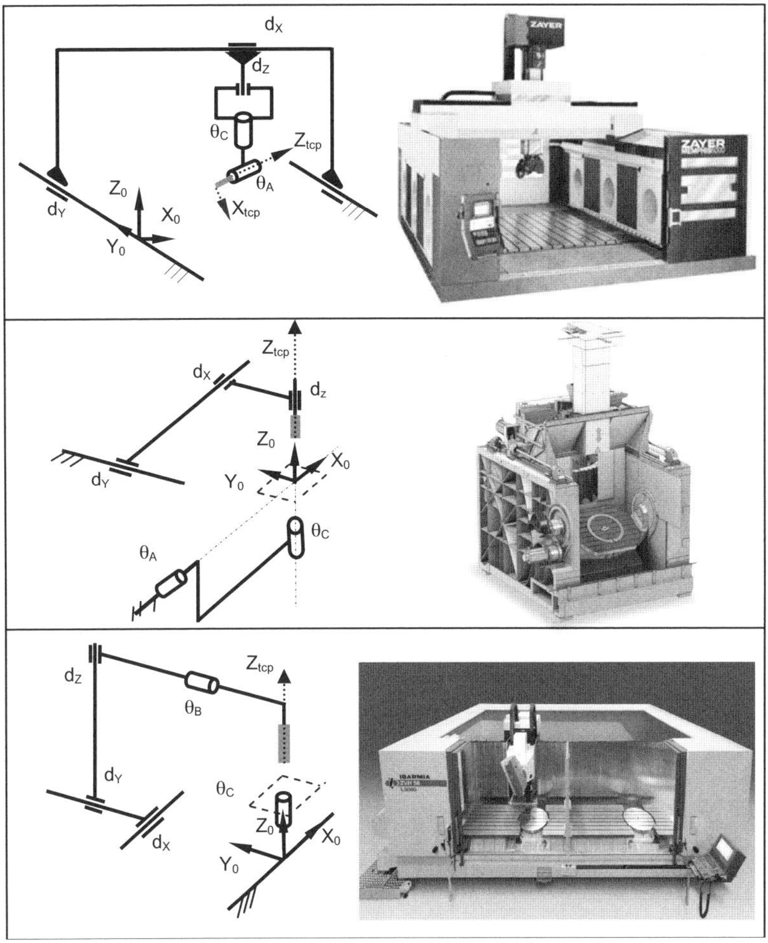

Figure 1.28. *Three mechanisms for a 5-axis HSM. Top, the LLLRR, Memphis model by Zayer®. Center, the RRLLL, Hermle C50 U model. Bottom, RLLLR, Ibarmia® ZVH55 with two work zones*

1.4.7.4 *Multitask machines*

A *multitask machine* allows different machining operations to be performed on the same machine (turning, milling, hobbing, etc.). Some now include grinding wheels and even laser heads. The main advantage of these machines is that all operations are performed on the same machine, using the same zero reference for the work piece, with no idle times between operations. There are several types of multitask machines, but the most common types are as follows:

– a horizontal lathe where a milling headstock is placed instead of the turret, or in addition to the turret. The movement of the C-axis can be held in position, being interpolated with the movements of the milling head. Almost all large manufacturers of lathes include these models in their catalogs;

– a vertical lathe including a milling ram. A turning tool can be placed in the ram, the rotation being blocked, or a milling tool can be rotated at high power and speed. In other configurations there are two separated rams, one including the spindle for milling and another without a spindle for turning;

– a 5-axis vertical machining center where the plate is replaced by a turning headstock. The motion of the headstock is blocked when milling;

– a new machine designed according to the end customer requirements, as is the case in the milling/turning machine shown in Figure 1.29.

a) b)

Figure 1.29. *The multitask machine Mori Seiki® Series NT: a) box-in-box structure with an octagonal section ram, b) performing a milling operation on a cylindrical work piece [MOR 08]*

1.4.8. *Cutting tool technology*

The most general and "first-choice" tool for HSM is the integral endmill made of *hard metal*, i.e. tungsten carbide micrograins sintered with a cobalt binder. Two aspects of this tool material are important: the amount of cobalt binder (6-12%) and the grain size. This means the terms micrograin (less than 1 μm) and submicrograin (less than 0.5 μm) are used – the smaller the grain, the harder and tougher the material. If the diameter is larger than 20 mm, a plate with carbide inserts is the most usual tool.

On the other hand, coatings for carbide tools are constantly evolving, increasing friction resistance and producing harder and chemically stable grades. In terms of milling coatings, the physical vapor deposition (PVD) technique is now constantly improving its performances, more so than chemical vapor deposition (CVD), and the latter is more generally used for turning. TiAlN and AlCrN coatings are "top notch" in coating technology today.

Extra-hard materials such as polycrystalline diamond (PCD) and polycrystalline cubic boron nitride (PcBN) are more expensive, but can be applied to significantly higher cutting speeds than carbide tools. PCD is a good solution for high-silicon alloys or abrasive composites, whereas PcBN is suitable for tempered steels and cast irons. However, these materials are more fragile than carbide and require a careful checking of the CNC programs. Even a gentle tool impact against the work piece may cause irreparable damage to the tool edges, and it is for this reason that industrial users prefer the tougher carbide tools.

1.5. High-throughput drilling

Drilling is a widely used machining process among all the machining processes. Drilling is usually used in the final steps of the mechanical component, manufacturing which involves high economical importance. Material deformation and tool geometry in throughput drilling are complex and difficult to study. A chisel edge located in the drill center has a cutting speed value close to zero and chip formation is machined under negative rake angle. Both rake angle and cutting speed vary with respect to the distance from the drill center along the drill cutting edge which performs differently from other machining processes. The external point on the cutting edge has the highest cutting speed, known as peripheral cutting speed. Figure 1.30 shows two types of cutting edge performance with internal coolant (see section 1.5.1.2) supply on drilling processes.

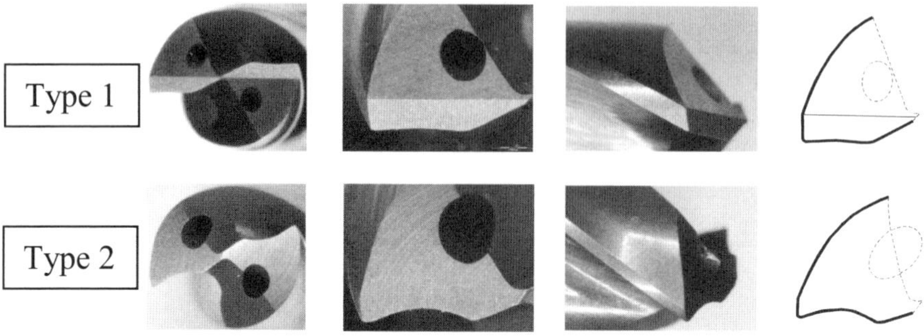

Figure 1.30. *Overview of tool geometry performance [GAN 08]*

Throughput drilling machine processes has some technical difficulties like:

1. chip ejection from the machining point;
2. hard drill wear;
3. limited cutting speed which means low productivity;
4. low dimensional and geometrical tolerances;
5. high drill temperature.

The material removal rate in throughput drilling processes is very low considering the effect of these technical restrictions. Although drilling has been utilized in industry, research studies are still limited. In general those research works analyze the effects of cutting speed, feed rate, use of cutting fluid on thrust force, torque, chip formation and tool wear in drilling processes. More advanced studies work on high-throughput drilling [SAK 96, ARA 97].

The main purpose of the drilling process is to increase productivity. In order to reach this goal, the material removal rate should be increased by increasing the feed rate and cutting speed known as high-throughput drilling. Using high cutting speed is the most common way to increase productivity [RAH 03]. The feed rate effect on high-throughput drilling is an area that has not been studied extensively. Tool geometry has an important influence on machine process performance. Otherwise, supplying cutting fluid to the cutting edge is usually needed to reduce the temperature. Appropriate drill geometry, cutting speed, feed rate and the cutting fluid supply method are important factors in achieving the high productivity in high-throughput drilling processes.

In throughput drilling processes, the thermal analysis of the tool is a critical factor to be analyzed due to the high temperatures conducted on the tool. This has a great effect on drill life. Considering the advances made on the materials and geometry of the drills lasting recent years, it is important to study the heat generation and temperature distribution when throughput drilling is carried out. In the thermal modeling of the throughput drilling process, the heat border between the chip and drill is proposed as one of the most difficult parameters to determine [AGA 94]. In order to estimate the heat border on the tool-chip contact area the inverse heat transfer method can be used. For throughput drilling, highly localized stresses and severe deformation are expected in the drill and will eventually lead to drill failure. The mechanical stress added to the thermal stress induced by temperature causes cutting forces to be studied to predict drill failure location.

A better understanding of the underlying mechanism of the throughput drilling process needs to be achieved by means of better process and performance models, thermal modeling and finite element simulation to obtain further reduction on the effective cost of machining. The process parameters influence the product quality and the capability to provide reliable predictions of the performance of the cutting process is demonstrated by the advancements in machining process modeling [EHM 97]. One extended model used in drilling processes is the updated-Lagrangian method. However, it has not yet been applied to throughput drilling due to the difficulty in modeling the material deformation under the complicated drill geometry, especially the large negative rake angle at the chisel edge used in throughput drilling processes.

In high-throughput drilling, the work-piece and chips have large deformation at high temperature and strain rate which alters the material properties and microstructure of the work-material and chips.

1.5.1. *Throughput drilling process parameters*

High cutting speed and feed per drill revolution are required in high-throughput drilling to achieve the desired MRR, cost and quality [STE 06].

In this section, the process parameters, tool material and drill geometry to achieve the high-throughput drilling is presented.

1.5.1.1. *Cutting and feed speed in throughput drilling*

Cutting and feed speed are two important process parameters to achieve the desired productivity and MRR in drilling. Larger cutting and feed speed in throughput drilling can be achieved when higher strength and hot hardness tool material and drill geometry is utilized.

The increase of cutting speed in drilling helps to improve the MRR and productivity. Depending on the tool and work materials, the cutting speed in drilling could vary considerably [SCH 92]. Aluminum alloys can be machined at 300 m/min using high-speed steel (HSS) and 600 m/min using tungsten carbide in cobalt matrix (WC-Co) tool materials [TRE 00]. For gray cast iron, typical cutting speeds could be from 40 up to 450 m/min, using HSS, WC-Co and ceramic cutting tools, respectively.

Besides cutting speed, the effect of feed in drilling also improves MRR and productivity. Feed speed in drilling processes is an area that has not been studied extensively, so it is becoming a high research interest. The feed of aluminum alloys can be elected from 1.00 to 2.05 mm/rev using HSS for lowers values and WC-Co tools for higher values. Gray cast iron can be machined at 0.65 up to 1.00 mm/rev with HSS, ceramic and WC-Co tools. New materials like titanium alloys are recommended to be machined with lower feeds of 0.40 mm/rev when WC-Co tools are used. The selection of the feed depends on the drill diameter. Sometimes in throughput drilling conditions, the adequate selection of feed is difficult because of the inherent work material properties, particularly for those materials with low thermal conductivity, which results in high tool temperature at increased cutting speed. High tool temperature accelerates tool wear and limits drill life [LI 06]. Table 1.2 shows some recommended values for cutting and feed speed in throughput drilling processes when using a WC-Co twist drill with diameter less than 6 mm.

	Cutting speed [m/min] (With internal cutting fluid supply)	Feed speed [mm/rev]
Aluminum alloys	105 (245)	0.18
Gray cast iron	49 (90)	0.10
Titanium alloys	11 (30)	0.05

Table 1.2. *Cutting and feed speed values using a WC-Co twist drill with diameter less than 6 mm [MAC 80]*

Currently, throughput drilling is utilized in industrial applications for Ti machining work parts due to high resistance and low weight of titanium alloys. It is the aeronautical industries which increase high-throughput drilling use. Machining of titanium alloys at high cutting speed has been studied by many researchers. In turning processes, Lei and Liu [LEI 02] conducted 480 m/min cutting speed at 0.15 mm/rev feed using a WC-Co tool. Zareena *et al.* [ZAR 01] achieved a 400 m/min cutting speed in face milling and 0.1 mm/tooth feed using a PcBN tool insert. However, all these studies were limited in either turning or milling and there are not many studies focused on drilling processes. However, Lin and Ni [LI 99] studied high-throughput drilling and they reached 533 and 235 m/min cutting speed and 0.50 and 0.36 mm/rev feed, on cast aluminum and gray cast iron, respectively, by using coated WC-Co specially designed drill geometry.

1.5.1.2. *Coolant techniques*

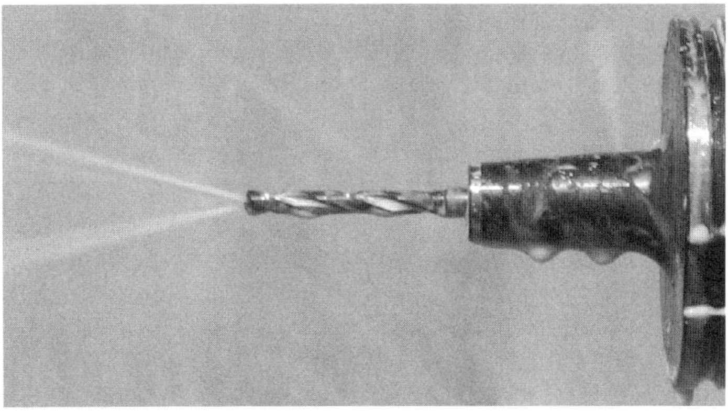

Figure 1.31. *High pressure coolant through machine spindle and drilling tool*

Cutting fluid use is also important to increase MRR and productivity and can be considered as a third parameter in throughput drilling processes. Both the cutting fluid and delivery method parameters are important to enhance the drill life and to reduce the tool and work-piece temperatures [TRE 00]. There are two ways to supply the cutting fluid: internally from two holes through the drill (Figure 1.31) and externally from outside the drill. The first way significantly increases the drill life due to effectiveness of lubricant and cooling effect on cutting edge and tool chip interface. This internal supply fluid also helps the chip ejection.

Some throughput drilling processes could apply high pressure up to 60 bars using emulsion coolant (4-5% of mineral oil in water). The internal technique tries to make better use of coolant than external flooding. Using internal coolant with high pressure reaches a large number of holes with the same drill, oil consumption and cost of lubricants. Future applications trend towards minimum quantity of lubricant (MQL) which implies decreasing the number of holes processed with the same process conditions but the consumption of biodegradable oil is significantly reduced. Coolant techniques should then be evaluated taking into account the global economic constrictions of the process for each company, application and environmental restrictions or requirements [GAN 08]. However, MQL does not break chips, which cannot be properly evacuated causing drill to break, mainly due to the higher torque. Figure 1.32 shows chip size/form for both internal and MQL lubricating systems.

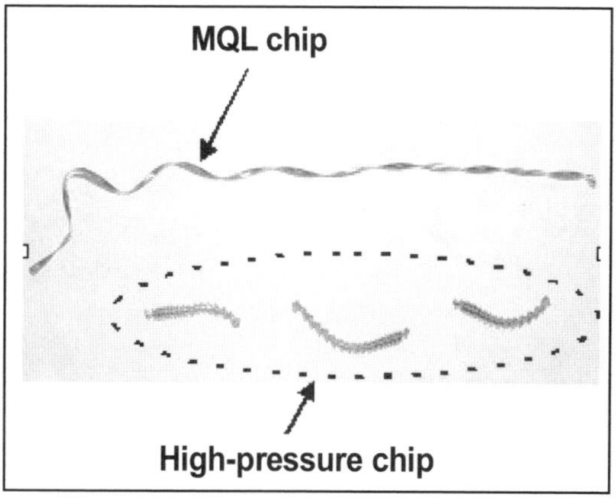

Figure 1.32. *Chips in high internal pressure and MQL tests [GRA 08]*

1.5.2. *Models for cutting forces in throughput drilling*

There are several models for cutting processes when considering turning or milling processes but it is difficult to find models for cutting forces in drilling or throughput drilling processes cases. Usually, models for cutting forces used in drilling processes are taken from turning or milling within some hypothesis or consideration. Using twist drills, which is the common case in drilling processes, the rake angle is negative and the chisel edge has more than one point under different conditions of cutting speed. With these conditions, it is difficult to study and obtain clear models for cutting forces like in turning.

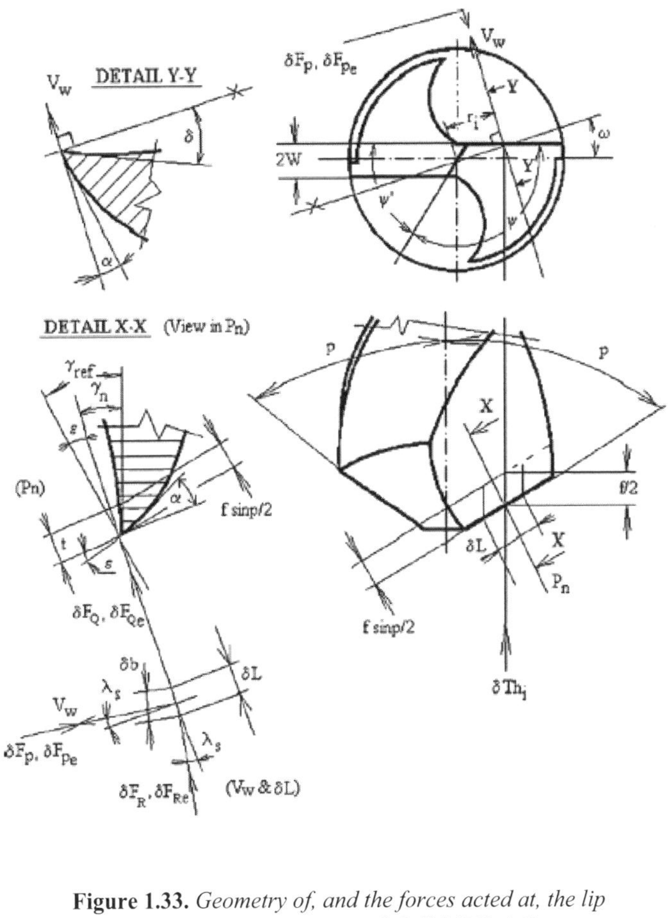

Figure 1.33. *Geometry of, and the forces acted at, the lip regions of a conventional drill [AUD 96]*

One good model for a cutting forces study is based on considering the chisel and cutting edges of a spiral point drill as a series of straight cutting edges like elementary cutting tools with various inclination and rake angles [LI 07a]. This consideration is combined with the oblique cutting model of each elementary cutting tool.

Figure 1.33 shows the geometry and the forces acting on an element at the lip region of a conventional general purpose twist drill. From the mathematical point of view, the drill lip region can be considered as a sum of a number of different elemental oblique cuts (see section 1.5.2.1) controlled principally by their own tool geometry and cutting conditions.

Other studies have analyzed drill tool and drilling processes under finite element overview since the 1970s [SAX 71, WAT 97]. Recently, for the drill temperature analysis, Bono and Ni [BON 06] have also applied the finite element method. All these research works have some limitations in giving accurate drill temperature prediction. Li and Shih [LI 07b] have again improved in their work, the finite element modeler applying the inverse heat transfer theory.

Criterion failure in high-throughput drilling can be analyzed with the three most frequently used criteria: Rankine, Mohr-Coulomb and modified Mohr criteria [BAR 01].

1.5.2.1. *Modeling cutting forces with elementary cutting tool*

An elementary cutting tool (ECT) [WAT 85, WAN 94] was used to divide the drill cutting edges into segments. By using this idea, Ke [KE 05] has developed an approach to extract the data forces working on each ECT, which again runs like an oblique cutting tool during drilling (Figure 1.34). At the start of drilling, one by one, each ECT in the drill cutting and chisel edges gradually engages into the cutting action from the center drill point. For each ECT, cutting forces can be calculated if thrust force and torque at the start of drilling process are known. From the machined chip the shear angle and chip thickness associated with each ECT to estimate the chip speed can be measured. Taking this cutting data, the heat generation and frictional force can be determined. The heat transferred between chip formation and drill can be adjusted by a heat partition factor in thermal finite element analysis. This heat partition factor between the drill and chip is recognized as an important and difficult parameter to determine.

Research studies use different alternatives for the heat partition factor consideration such as constant heat partition factor value along the tool-chip interface [AGA 94, SAX 71, WAT 77] and variable heat partition factor depending on the cutting speed [BER 91], also adopted for drilling temperature prediction by Bono and Ni [BON 06]. On 2D orthogonal cutting considerations, more recent and advanced modeling variable heat partition factors have been developed in the sticking and sliding contact regions [HUA 05b, KAR 05].

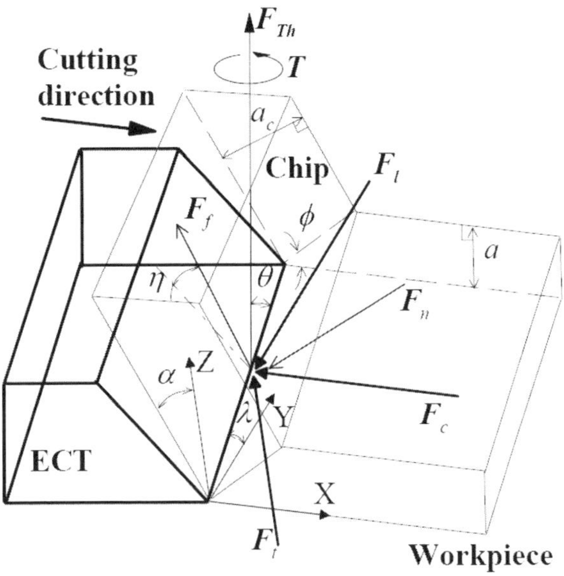

Figure 1.34. *Elementary cutting tool (ECT) [LI 07]*

Figure 1.34 shows the oblique cutting model of an ECT model with five angles represented: normal rake angle α, chip flow angle η, shear angle φ, inclination angle λ and the angle between the drill axis and ECT cutting edge θ. Chip flow angle η is considered to be equal to inclination angle λ. The chip thickness and uncut chip thickness are shown as *a* and a_c, respectively.

Figure 1.34 uses orthogonal coordinate systems with X-axis in the cutting direction and Z-axis perpendicular to the plane determined by the straight cutting edge and X-axis. Finally, the Y-axis is perpendicular to the X and

Z-axes. The torque (T) generates along the X-axis direction a force component F_c. F_c = T/r where r is the distance from the drill axis to the center of the ECT, Fl is the component of resultant force in the Y-axis and Ft is the same but in the Z axis. When F_t and F_l are composed parallel to the drill-axis, we obtain the thrust force, (F_{Th}) [BON 06].

$$F_{Th} = -F_l \frac{\cos\theta}{\cos\lambda} + F_t \frac{\sqrt{\cos^2\lambda - \cos^2\theta}}{\cos\lambda} \quad [1.1]$$

The resultant force on an ECT can also be decomposed to the force components normal and parallel to the rake face, denoted as F_n and F_f. Friction force in the direction of chip flow on the tool rake face is represented by F_f. As the resultant force is included in the plane defined by F_n and F_f, F_l is related to F_c and F_t by [LIN 82]:

$$F_l = \frac{F_c(\sin\lambda - \cos\lambda\sin\alpha\tan\eta) - F_t\cos\alpha\tan\eta}{\sin\lambda\sin\alpha\tan\eta - \cos\lambda} \quad [1.2]$$

F_c, F_l, and F_t are related to F_n and F_f by:

$$\begin{Bmatrix} F_l \\ F_c \\ F_t \end{Bmatrix} = \begin{bmatrix} \cos\lambda & \sin\lambda & 0 \\ \sin\lambda & \cos\lambda & 0 \\ 0 & 0 & 1 \end{bmatrix} \begin{bmatrix} -1 & 0 & 0 \\ 0 & \sin\alpha & \cos\alpha \\ 0 & \cos\alpha & \sin\alpha \end{bmatrix} \begin{bmatrix} 0 & \sin\eta \\ 0 & \cos\eta \\ 1 & 0 \end{bmatrix} \begin{Bmatrix} F_n \\ F_f \end{Bmatrix} \quad [1.3]$$

F_f can be calculated as:

$$F_f = \frac{(\cos\alpha\cos\lambda)F_t - \sin\alpha * F_c}{\cos\lambda\cos\eta + \sin\alpha\sin\lambda\sin\eta} \quad [1.4]$$

The F_t can be solved by substituting equation (1.2) into (1.1).

$$F_t = \frac{F_{Th}(\sin\lambda\sin\alpha\tan\eta + \cos\lambda)\cos\lambda + F_c(\sin\lambda - \cos\lambda\sin\alpha\tan\eta)\cos\theta}{\sqrt{\cos^2\lambda - \cos^2\theta}\,(\sin\lambda\sin\alpha\tan\eta + \cos\lambda) + \cos\alpha\tan\eta\cos\theta} \quad [1.5]$$

The chip velocity V_c along the chip flow direction could also relate to the frictional heat generation. Considering ω as the drill rotational speed, cutting speed V_c could also be calculated with $V_c = \omega r$ [SHA 84].

$$V_c = V \frac{(\cos \lambda \sin \phi)}{\cos \eta \cos (\phi - \alpha)} \qquad [1.6]$$

The chip thickness a_c corresponding to each ECT could experimentally be measured and used to calculate the shear angle φ [SHA 84]:

$$\tan \phi = \frac{a \cos \alpha}{a_c - a \sin \alpha} \qquad [1.7]$$

where

$$a = f_d \sin \theta$$

in which f_d is the feed per tooth and θ is the angle between the drill axis and ECT cutting edge.

1.5.2.2. *Modeling cutting forces with finite element modeler*

Difficulty in modeling the large negative rake angle cutting which occurs in the drill center (at the chisel edge) and complexity of tool geometry means that the research in the FEM of drilling is limited compared to numerous models of orthogonal and oblique cutting. Strenkowski *et al.* [STR 04] used the Eulerian formulation to predict the forces and torque in drilling for each ECT with the oblique cutting model by dividing conventional twist drill into a series of ECT approaches.

Based on twist drill, Min *et al.* [MIN 01] developed a 2D burr formation model and Guo and Dornfeld [GUO 00] investigated burr formation on a 3D drilling process both of them with the updated-Lagrangian method. Last but not least, Marusich *et al.* [MAR 06] also simulated, with conventional twist drill, a 3D drilling process using Ti-6Al-4V material.

However, the updated-Lagrangian method, due to the technical difficulty in modeling the material deformation under the large negative rake angle at the chisel edge, has not been applied to predict the thrust force and torque either in drilling or throughput drilling processes.

1.5.2.3. *Modeling temperature distribution in throughput drilling*

For throughput drilling, high tool temperature softens the tool material and is detrimental to the tool life. Since the tool is constrained in a hole, the high temperature is significant in the drill tip. This results in the limited material removal rate (MRR) and, subsequently, low productivity and high cost in machining.

The drill temperature and cutting forces can be used as inputs to analyze the spatial and temporal distributions of the drill stress. For high-throughput drilling, the drill temperatures and forces are both very high. It is known that the supply of cutting fluid and the selection of feed and cutting speed in drilling greatly affect drill temperature, stress, and tool life. Severe deformation and highly localized stresses are expected in the drill and will eventually lead to the drill failure when using the throughput drilling process.

Several methods such as using an embedded foil tool-work thermocouple, micro-hardness measurement, scanning electron microscopy, chemical composition change or even metallurgical transformation and energy dispersive x-ray have been developed to measure the drill temperature distribution in drilling processes. However, in throughput drilling processes, spatial and temporal drill temperature distribution using the inverse heat transfer method seems to be the best method to obtain good predictions and results.

The drill modeling method used in the inverse heat transfer method is the same as in section 1.5.2.1. Cutting edge and chisel edge is represented by ECTs. A cutting analysis was conducted to calculate the heat generation rate (q_{tool}) in each ECT. In order to convert the heat generation rate by friction on the tool–chip interface q_f to q_{tool} heat partition factor K is utilized [LI 07b]:

$$q_{tool} = q_f K \qquad [1.8]$$

where q_f is the product of the chip velocity V_c and the friction force F_f. Measuring thrust force, torque, and chip thickness on each ECT allows us to obtain V_c and F_f. The heat partition factor K can be calculated as equation [1.9] shows and it determines the ratio of heat transferred to the tool:

$$K = 1 - \left(1 + 0.45 \frac{k_t}{k_w} \sqrt{\frac{\pi d_w}{V_c l}}\right)^{-1} \qquad [1.9]$$

where k_t and k_w are the thermal conductivities of tool and work piece material, respectively, d_w is the work piece diffusivity and l is the tool–chip contact length. All thermal properties are temperature-dependent.

The shear angle and chip thickness associated with each ECT are measured from the machined chip to estimate the cutting speed V_c. Frictional force and heat generation rate q_f can be determined with information obtained in each ECT. The heat partition factor K is required to find the amount of heat transferred to the drill q_{tool} for thermal element analysis. By assuming an initial value for s (number of ECT elements on cutting edge), the values of l, K and q_{tool} can be calculated and applied to ECT nodes on the chisel and cutting edges. The spatial and temporal temperature distribution of the drill can then be solved.

1.5.3. *Optimal conditions in throughput drilling*

Optimizing drilling process conditions has been researched in many innovative works during recent years. In this section, some general steps towards selecting optimal throughput drilling conditions are presented.

Figure 1.35 shows the main tool wear effects carried out in throughput drilling processes. The first image shows the cutting edge where it intersects with maximum diameter where the cutting speed has the highest value, and the second in the chisel edge for a three cutting edge drill. Tool wear in the chisel edge is carried out due to the penetration effect when the drill advances because there is no cutting effect on the drill center where cutting speed is zero, and thus the drill penetrates without cutting.

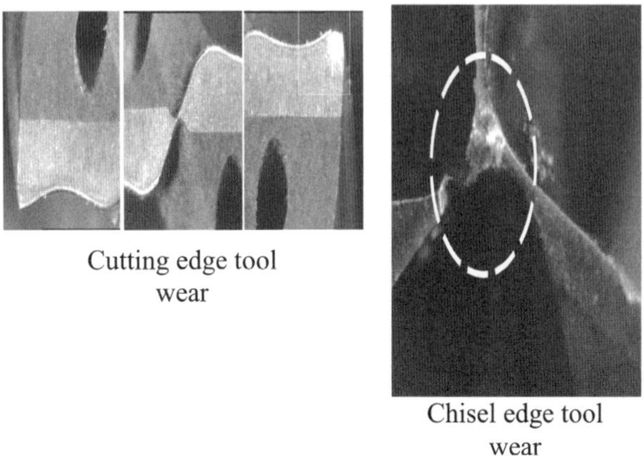

Figure 1.35. *Tool wear in drill tools*

Research works [LOP 08] demonstrate how power trust in drilling processes increases when tool wear also increases. However, this relationship is not linear, as can be observed in Figure 1.36. This figure shows the trend of power consumptions by the number of holes manufactured. Power consumptions are measured by electrical intensity on the main spindle where both parameters are linear-related. Measuring the intensity is easy and power consumption is a good criterion to have in throughput drilling processes.

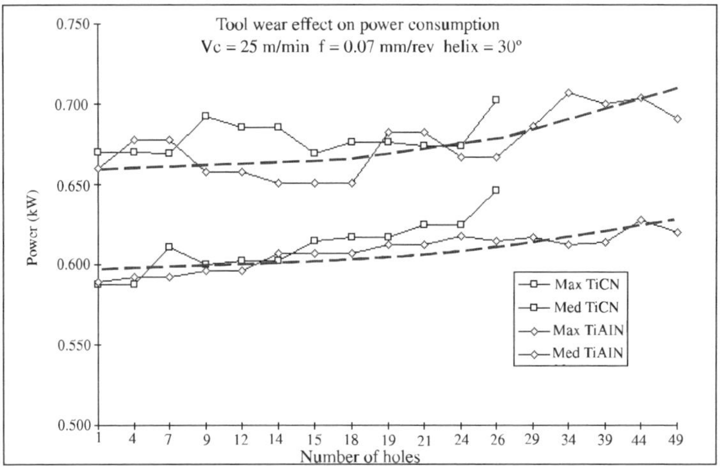

Figure 1.36. *Tool wear effect on power consumption*

Usual intensity value trends are shown in Figure 1.37. Intensity value increases when the drill tool starts to penetrate into the work piece and afterwards it keeps its value during the entire throughput drilling process to finally decrease when the drill tool passes over the other side of the work piece. When a hole is manufactured, the intensity value increases slowly due to tool friction in the back path.

Trends on throughput drilling processes are always similar to those shown in Figures 1.36 and 1.37. More tool wear and worse cutting edge performance on drill tool cutting operation will have worse quality and precision. Furthermore, when tool wear increases, power consumption also increases and hole position and diameter could be more difficult to obtain with quality values.

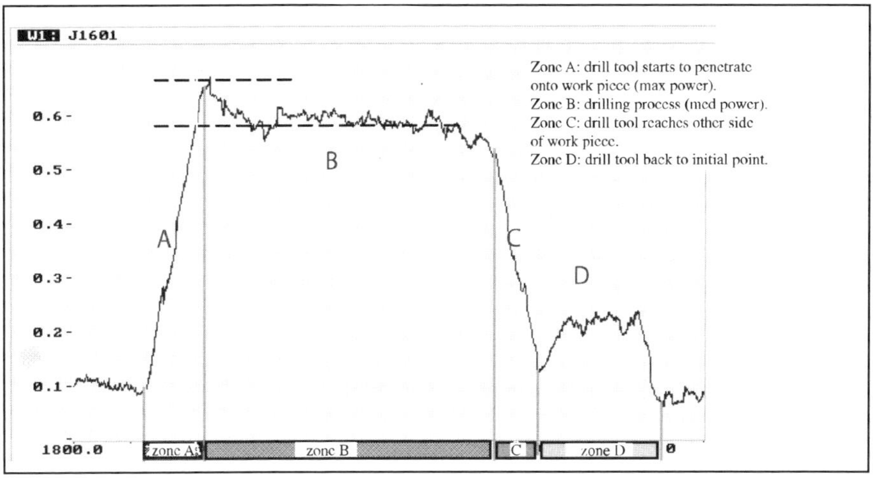

Figure 1.37. *Intensity trend when one hole is manufactured*

It is also important to avoid burr formation either on entrance or exit of drill tool from the work piece. Burr formation obliges us to perform one more task to reduce it, which means an additional operation and more time to carry out the whole drilling process. Two processes for one hole should be planned for automatic operations. Figure 1.38 shows three types of burr formation for throughput drilling processes. Burr formation could be avoided by controlling tool wear, good drill tool design and good feed rate selection.

Figure 1.38. *Burr formation in drilling processes (Courtesy of the Consortium on Deburring and Edge Finishing (CODEF), University of California, Berkeley, USA)*

1.6. Environmentally benign manufacturing

Manufacturing with minimum impact on the environment, known as Environmentally Benign Manufacturing (EBM) or Green Manufacturing [GUT 05], emerged as the result of two major factors. Firstly, intense social pressure in connection with sustainable development that prevents pollution, with the direct result of tighter legislation to take action against those who pollute or waste resources, and an increasing tendency towards charges for pollution-related activity. Secondly, savings and environmental efficiency of processes can significantly lower production costs.

A number of different pieces of software are frequently used in industrialized nations to carry out life-cycle analyses (LCA). LCAs are a tool to assess the potential impact of a process, an activity or a product on the environment throughout their entire life-cycles, by quantifying the utilization of resources such as energy, raw materials and water and environmental emissions, i.e. the impact on soil, air or water. LCAs are extremely important tools in relation to implementation of the ISO 14000 environmental management standards initiated in 1996 (ISO 14001:2004 environmental management systems – requirements with guidance for use).

From a technological point of view, there have been two main lines of development in terms of green machining: reduction or total elimination of coolants, and reduction of machine power consumption.

1.6.1. *Dry and near-to-dry machining*

Dry machining implies the total elimination of coolants, oils or emulsions. Elimination of liquid coolants or lubricants can produce savings of up to 10-15% on total production costs [KLO 97]. The feasibility of dry machining, however, depends on the material to be machined:

– in the case of tempered steels, dry machining is highly recommended using a flow of air to remove chips;

– in the case of routing on Carbon Fiber Reinforced Plastics (CFRP) or Glass Fiber Reinforced Plastics (GFRP), dry high-speed milling is used taking due care to use dust suction systems;

– in the case of light alloys (aluminum), a solution has not yet been found for totally dry machining, and a near-to-dry approach has been taken using the Minimum Quantity of Lubricant (MQL) technique [SRE 08, LOP 04]. The MQL system operates with compressed air, injecting a mixture of air and oil mist into the cutting zone. As the example in Figure 1.24 shows, during the milling of aluminum alloys oil consumption (biodegradable ester) can be reduced to 0.04 cc/min by applying cutting conditions producing a chip flow rate of 45,000 mm^3/min. This means that 420 hours of machining may be secured from one liter of oil thereby eliminating 1.2 m^3 (3 tons) of aluminum;

– in the case of cast iron or aluminum cast alloys for automotive components, MQL is an extremely common practice in view of the need for constant cost reductions in mass production processes;

– in the case of stainless steels, good results with MQL have been obtained in experimental research work [AOY 08, GAN 08], but MQL is not yet commonly used;

– dry machining is still unfeasible in the case of nickel, cobalt or titanium alloys, and MQL is rarely practical apart from certain research projects.

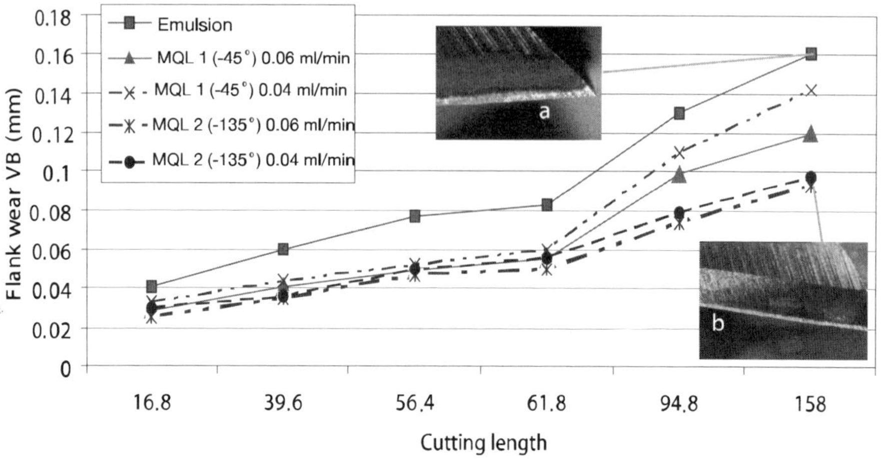

Figure 1.39. *The trend in flank wear, with respect to milling length in experimental tests, maximum wear in the case of emulsion coolant and minimum wear in a consumption of 0.06 cc/min to 135° (just ahead of the milling tool, 45° is just behind it), a) wear with the emulsion coolant after machining 158 m, b) wear with 0.04 cc/min MQL to 135° after machining 158 m*

1.6.2. *Reduction of machine power consumption*

A slight reduction in the power consumption of a machine tool entails constant year-on-year savings. Over the last five years several development projects (the EU's integrated project NEXT, for example) have focused on reducing power consumption, leading to the manufacture of "green machines". The objective of NEXT [NEX 08] is to produce machines that consider environmental aspects over their entire life-cycle: use of recycled materials for machine elements (>50%), reduction of energy consumption (by at least 25%) during machine operation, zero waste produced, dismantling and recycling of 100% of machines.

The electrical energy that a machine consumes is largely associated with the motion of moving structural components both in positioning motions and in machining motions. The reduction of each gram of material in these movable structural components, will thus play a decisive role in the final ecological impact of the machine tool. All reductions in mass must nevertheless be considered from the point of view of the required machine static and dynamic stiffness.

1.7. References

[AER 00] AEROSPACE ED, 2000, "Milling machine for Airbus Wings", *Aircraft Engineering and Aerospace Technology*, vol. 72, no. 6.

[AER 02] AEROSPACE ED., 2003, "High speed milling", *Aircraft Engineering and Aerospace Technology*, vol. 75, no. 4.

[AGA 94] AGAPIOU J. S., STEPHENSON D. A., 1994, "Analytical and experimental studies of drill temperatures", *Journal of Engineering for Industry*, 116(1), pp. 54-60.

[AOY 08] AOYAMA T., KAKINUMA Y., YAMASHITA, M., AOKI, M., 2008, "Development of a new lean lubrication system for near dry machining process", *Annals of the CIRP*, vol. 57, no. 1, 2008, pp. 125-128.

[ARA 97] ARAI M., OGAWA M., 1997, "Effects of high pressure supply of cutting fluid in drilling of titanium alloy", *Keikinzoku/Journal of Japan Institute of Light Metals*, 47(3), pp. 139-144.

[ARA 08] ARRAZOLA P., ÖZEL T., 2008, "Numerical modelling of 3D hard turning using the arbitrary Eulerian Lagrangian finite element method", *International Journal of Machining and Machinability of Materials*, vol. 3, no. 3, pp. 238-249.

[AUD 08] AUDY J., 2008, "A study of computer-assisted analysis of effects of drill geometry and surface coating on forces and power in drilling", *Journal of Processing Technology*, 204, pp. 130-138.

[AUR 06] AURICH J.C., BIL H., 2006, "3D finite element modelling of segmented chip formation", *Annals of the CIRP*, 55/1: 47-50.

[BAR 01] BARBER J. R., 2001, *Intermediate Mechanics of Materials*, McGraw-Hill, Boston.

[BER 91] BERLINER E. M., KRAINOV V. P., 1991, "Analytic calculations of the temperature field and heat flows on the tool surface in metal cutting due to sliding friction", *Wear*, 143, pp. 379-395.

[BON 06] BONO M., NI J., 2006, "The location of the maximum temperature on the cutting edges of a drill", *International Journal of Machine Tools and Manufacture*, 46, 901-907.

[BRA 05] BRAVO U. ALTUZARRA O., LÓPEZ DE LACALLE L.N., SÁNCHEZ J.A., CAMPA F.J., 2005, "Stability limits of milling considering the flexibility of the work piece and the machine", *International Journal of Machine Tools and Manufacture*, vol. 45, no. 15, pp. 1669-1680.

[BRI 97] BRINKSMEIER E., BERGER U., JANSSEN R., "High speed milling of Ti-6Al-4V for aircraft applications", *First French and German Conference on High Speed Machining*, Metz, June 1997.

[BUD 80] BUDINSKI K.G., "Tool materials", *Wear Control Handbook*, ASME, 1980, New York.

[BYR 03] BYRNE G., DORNFELD D., DENKENA B., 2003, "Advancing cutting technology", *CIRP Annals – Manufacturing Technology*, vol. 52, no. 2, pp. 483-507.

[CAM 07] CAMPA F., LÓPEZ DE LACALLE L.N, LAMIKIZ A., SÁNCHEZ J.A., 2007, "Selection of cutting conditions for stable milling of flexible parts with bull-nose end mills", *Journal of Materials Processing Technology*, vol. 191, no. 1-3, pp. 279-282.

[CHE 06] CHEN L., ELWARDANY T.I., NASR M., ELBESTAWI M.A., 2006, "Effects of edge preparation and feed when hard turning a hot work die steel with polycrystalline cubic boron nitride tools", *Annals of the CIRP*, 55/1: 88-92.

[COE 07] COELHO R.T., NG E., ELBESTAWI M.A., 2007, "Tool wear when turning hardened AISI 4340 with coated PcBN tools using finishing cutting conditions", *International Journal of Machine Tools and Manufacture*, 47: 263-272.

[DAV 96] DAVIES M.A., CHOU Y., EVANS C.J., 1996, "On chip morphology, tool wear and cutting mechanics in finish hard turning", *Annals of the CIRP*, vol. 45/1, pp.77-82.

[DAV 97] DAVIES M.A., BURNS T.J., EVANS C.J., 1997, "On the dynamics of chip formation in machining hard metals", *Annals of the CIRP*, 46/1: 25-28.

[DEN 05] DENKENA B., BECKER J.C., DE LEON-GARCIA L., 2005, "Study of the influence of the cutting edge microgeometry on the cutting forces and wear behavior in turning operations", *Proc. 8th CIRP, Int. Workshop on Modelling of Machining Operations*, Chemnitz, Germany, 503-507.

[DEW 97] DEWES R.C., ASPINWALL D.K., "A review of ultra high speed milling of hardened steels", *Journal of Materials Processing Technology*, vol. 69, 1997, pp. 1-17.

[EHM 97] EHMANN K.F., KAPOOR S. G., DEVOR R. E., LAZOGLU I., 1997, "Machining process modeling: a review", *Journal of Manufacturing Science and Engineering*, 119, pp. 655-663.

[ELB 96] ELBESTAWI M.A., SRIVASTAVA A.K., EL-WARDANY T.I., "A model for chip formation during machining of hardened steel", *Annals of the CIRP*, vol. 45/1, 1996, pp. 71-76.

[ELB 97] ELBASTAWI M.A., CHEN L., BECZE C.E., EL-WARDANY T.I., "High-speed milling of dies and molds in their hardened state", *Annals of the CIRP*, vol. 46/1, 1997, pp. 57-62.

[FAL 99] FALLBÖHMER P., RODRIGUEZ C., ÖZEL T., ALTAN T., 1999, "High speed machining of cast iron and alloy steels for die and mold manufacturing", *Journal of Materials Processing Technology*, vol. 98, 104-115.

[FLO 83] FLOM D. G., 1983, "GE Report No. 83-SRD-040", *Advanced Machining Research Program (AMRP)*, Oct. 15.

[FLO 84] FLOM D.G., KOMANDURI R., LEE M., 1984, "High-speed machining of metals", *Annual Review of Materials Science*, 1984, vol. 14, pp. 231-278.

[GAN 08] GANDARIAS A., LÓPEZ DE LACALLE L.N., AIZPITARTE X., LAMIKIZ A., 2008, "Study of the performance of the turning and drilling of austenitic stainless steels using two coolant techniques", *International Journal of Machining and Machinability of Materials*, vol. 3, no. 1-2, pp. 1-17.

[GUO 00] GUO Y. B., DORNFELD D. A., 2000, "Finite element modeling of burr formation process in drilling 304 stainless steel", *Journal of Manufacturing Science and Engineering*, 122, pp. 612-619.

[GUO 02] GUO Y., LIU C.R., 2002, "3D FEA modeling of hard turning", *ASME Journal of Manufacturing Science and Engineering*, 124: 189-199.

[GUT 05] GUTOWSKI T., MURPHY C., ALLEN D., BAUER D., BRAS B., PIWONKA T., SHENG P., SUTHERLAND J., THURSTONI D., WOLFF E., 2005, "Environmentally benign manufacturing: observations from Japan, Europe and the United States", *Journal of Cleaner Production*, vol. 13, no. 1, pp. 1-17.

[HER 05] HERRANZ S., CAMPA F.J., LÓPEZ DE LACALLE L.N. RIVERO A., LAMIKIZ A., 2005, "The milling of airframe components with low rigidity: a general approach to avoid static and dynamic problems", *Proceedings of the Institution of Mechanical Engineers, Part B: Journal of Engineering Manufacture*, vol. 219, no. 11, pp. 789-801.

[HOL 94] HOLMBERG K., MATTHEWS A., *Coatings and Tribology*, 1994, Elsevier Publications, Amsterdam.

[HUA 05a] HUA J., SHIVPURI R., CHENG X., BEDEKAR V., MATSUMOTO Y., HASHIMOTO F., WATKINS T.R., 2005, "Effect of feed rate, work piece hardness and cutting edge on subsurface residual stress in hard turning of bearing steel using chamfer and hone edge geometry", *Materials Science and Engineering*, A394: 238-248.

[HUA 05b] HUANG Y., and LIANG S. Y., 2005, "Cutting temperature modeling based on non-uniform heat intensity and partition ratio," *Machining Science and Technology*, 9(3), pp. 301-323.

[ISH 98] ISHEE D., 1998, A comparison of the distortion of thin aluminum ribs in high speed machining and low speed machining, Thesis, Mississippi State University.

[KAR 05] KARPAT Y., ÖZEL T., 2005, "An analytical-thermal modeling approach for predicting forces, stresses and temperatures in machining with worn tools", *Proceedings of 2005 ASME International Mechanical Engineering Congress and Exposition*, 16(1), pp. 489-498.

[KAR 07] KARPAT Y., ÖZEL T., 2007, "3-D FEA of hard turning: investigation of PCBN cutting tool micro-geometry effects", *Transactions of NAMRI/SME*, vol. 35, pp. 9-16.

[KAR 08a] KARPAT Y., ÖZEL T., 2008, "Mechanics of high speed machining with curvilinear tools", *International Journal of Machine Tools and Manufacture*, 49: 195-208.

[KAR 08b] KARPAT Y., ÖZEL T., 2008, "Analytical and thermal modeling of high-speed machining with chamfered tools", *ASME Journal of Manufacturing Science and Engineering*, 130: 1-15.

[KE 05] KE F., NI J., STEPHENSON D. A., 2005, "Continuous chip formation in drilling", *International Journal of Machine Tools and Manufacture*, 45(15), pp. 1652-1658.

[KLO 05a] KLOCKE F., BRINKSMEIER E., WEINERT K., 2005, "Capability profile of hard cutting and grinding processes", *Annals of the CIRP*, 54/2: 557-580.

[KLO 05b] KLOCKE F., KRATZ H., 2005, "Advanced tool edge geometry for high precision hard turning", *Annals of the CIRP*, 54/1: 47-50.

[KLO 97] KLOCKE F., EISENBLÄTTER G., 1997, "Dry cutting", *Annals of the CIRP*, vol. 46/2, pp. 1-8.

[KOM 81] KOMANDURI R. and HAZRA J., 1981, "A metallurgical investigation of chip morphology in machining and AISI 4045 steel at various speeds up to 10.100 sfpm", *9th NAMRC*, 1981, pp. 356-373.

[KOM 82] KOMANDURI R., 1982, "Some clarifications on the mechanics of chip formation when machining titanium alloys", *Wear*, 76, pp. 15-34.

[KÖN 1993] KÖNIG W., BERKTOLD A., KOCH K-F., "Turning versus grinding – a comparison of surface integrity aspects and attainable accuracies", *Annals of the CIRP*, vol. 42/1, 1993, pp. 39-43.

[KÖN 84] KÖNIG W., 1984, "Fertigungsverfaren", *VDI Verlag*, Düsseldorf, Germany.

[KOV 95] KOVACEVIC R., CHERUKUTHOTA C., MOHAN R., 1995, "Improving milling performance with high pressure waterjet assisted cooling", *J. of Eng. for Industry*, vol. 117, pp. 331-339.

[KRA 93] KRAMER B.M., VIENS D., CHIN S., 1993, "Theoretical consideration of rare earth metal compounds as tool materials for titanium alloys", *Annals of the CIRP*, vol. 42-1, pp 111-114.

[KRO 61] KRONENBERG M., 1961, "Gedanken zur Theorie und Praxis der Ultra chnellzerspanung", *TZ Metallbearbeitung*, 55.

[LEE 81] LEE M., and FLOM D. G., 1981, "Metallurgical aspects of the chip formation process at very high cutting speed", *9th NAMRC*, 1981, pp. 326-333.

[LEI 02] LEI S. and LIU W., 2002, "High-speed machining of titanium alloys using the driven rotary tool", *International Journal of Machine Tools and Manufacture*, 42(6), pp. 653-661.

[LI 06] LI R., RIESTER L., WATKINS T.R., BLAU P.J. and SHIH A. J., 2006, "Metallurgical analysis and nanoindentation characterization of Ti-6Al-4V work piece and chips in high throughput drilling", *Materials Science and Engineering: A, 472*, pp. 115-124.

[LI 07a] LI R., HEGDE P. and SHIH A.J., 2007, "High-throughput drilling of titanium alloys", *International Journal of Machine Tools and Manufacture*, 47, pp. 63-74.

[LI 07b] LI R. and SHIH A.J., 2007, "Spiral point drill temperature and stress in high-throughput drilling of titanium", *International Journal of Machine Tools and Manufacture*, 45, pp. 2005-2017.

[LIN 82] LIN G.C., MATHEW P., OXLEY P.L.B., WATSON A.R., 1982, "Predicting cutting forces for oblique machining conditions", *Proceedings of the Institution of Mechanical Engineers*, 196, pp. 141-148.

[LIN 99] LIN X. and NI J., 1999, "Drill design for high throughput hole making", 1999, *International Conference on Advanced Manufacturing Technology*, June 16-18, 1999, Xi'an, China, pp. 162-166.

[LOP 02a] LÓPEZ DE LACALLE L.N., LAMIKIZ A., SALGADO M.A., HERRANZ S., RIVERO A., 2002, "Process planning for reliable high-speed machining of moulds", *Int. Journal of Production Research*, vol. 40, no. 12, pp. 2789-2809.

[LOP 02b] LÓPEZ DE LACALLE L.N., LAMIKIZ A., SÁNCHEZ J.A., ARANA J.L., 2002, "Improving the surface finish in the high speed milling of stamping dies", *Journal of Materials Processing Technology*, vol. 123, no. 2, pp. 292-302.

[LOP 04] LÓPEZ DE LACALLE L.N., SÁNCHEZ J.A., LAMIKIZ A., CELAYA A., 2004, "Plasma-assisted milling of heat-resistant superalloys", *Journal of Manufacturing Science and Engineering*, vol. 126, no. 2, pp. 274-285.

[LOP 06] LÓPEZ DE LACALLE L.N., ANGULO C., LAMIKIZ A., SÁNCHEZ J.A., 2006, "Experimental and numerical investigation of the effect of spray cutting fluids in high speed milling", *Journal of Materials Processing Technology*, vol. 172, no.1, pp. 11-15.

[LOP 07] LÓPEZ DE LACALLE L.N., LAMIKIZ A., SÁNCHEZ J.A., SALGADO M.A., 2007, "Toolpath selection based on the minimum deflection cutting forces in the programming of complex surfaces milling", *International Journal of Machine Tools and Manufacture,* 47: 388-400.

[LOP 08] LÓPEZ DE LACALLE L.N., LAMIKIZ A. (eds), 2008, *Machine Tools for High Performance Machining*, Springer Verlag.

[MAC 80] MACHINABILITY DATA CENTER, *Machining Data Handbook*, Metcut Research Associated, Inc., 1980.

[MAR 95] MARUSICH T.D., ORTIZ M., 1995, "Finite element simulation of high speed machining, simulation of material processing", *Proceedings of the NUMIFORM'95*, pp. 101-108.

[MAR 06] MARUSICH T. D., USUI S., APHALE R., SAINI N., LI R., SHIH A. J., 2006, "Three-dimensional finite element modeling of drilling processes", *2006 ASME International Conference on Manufacturing Science and Engineering*, October 8-11, 2006, Ypsilanti, Michigan.

[MAT 99] MATSUMOTO Y., HASHIMOTO F., LAHOTI G., 1999, "Surface integrity generated by precision hard turning", *Annals of the CIRP*, 48/1, 59-62.

[MCG 78] MCGEE F.J., 1978, "Manufacturing methods for high-speed machining of aluminum". *Final Technical Report (DRDMI-EAT) Requirement no. 6089*, US Army Missile R & D Command.

[MIN 01] MIN S., DORNFELD D. A., KIM J., SHYU B., 2001, "Finite element modeling of burr formation in metal cutting", *Machining Science and Technology*, 5, pp. 307-322.

[MOO 75] MOORE, D.F., 1975, *Principles and Applications of Tribology*, Pergamon Press, UK.

[MOR 08] MORI SEIKI, www.moriseiki.com, accessed in September 2008.

[NAK 74] NAKAYAMA K., 1997, "The formation of saw tooth chips", *Proceeding of International Conference on Production Engineering*, Tokyo, pp. 572-579.

[NEX 08] NEXT, www.nextproject.eu, accessed in September 2008.

[NG 97] NG E.G., ASPINWALL D.K., 1997, "An analytical approach to the machining of AISI H13 hot work steel using PCBN tooling", *International Conference and Exhibition on Design and Production of Dies and Molds*, pp.123-130.

[ÖZE 03] ÖZEL T., 2003, "Modeling of hard part machining: effect of insert edge preparation for CBN cutting tools", *Journal of Materials Processing Technology*, vol. 141, pp. 284-293.

[ÖZE 05a] ÖZEL T., HSU T.-K., ZEREN E., 2005, "Effects of cutting edge geometry, work piece hardness, feed rate and cutting speed on surface roughness and forces in finish turning of hardened AISI H13 steel", *International Journal of Advanced Manufacturing Technology*, vol. 25, pp. 262-269.

[ÖZE 05b] ÖZEL T., KARPAT Y., 2005, "Predictive modeling of surface roughness and tool wear in hard turning using regression and neural networks", *International Journal of Machine Tools and Manufacture*, vol. 45, pp. 467-479.

[ÖZE 07] ÖZEL T., ZEREN E., 2007, "Finite element modeling of the influence of edge roundness on the stress and temperature fields induced by high speed machining", *International Journal of Advanced Manufacturing Technology*, vol. 35, no. 3-4, pp. 255-267.

[ÖZE 08] ÖZEL T., KARPAT Y., SRIVASTAVA A.K., 2008, "Hard turning with variable micro-geometry PcBN tools", *CIRP Annals – Manufacturing Technology*, 57, 73-76.

[POU 01] POULACHON G., MOISAN A., JAWAHIR I.S., 2001, "Tool-wear mechanisms in hard turning with polycrystalline cubic boron nitride tools", *Wear*, 250: 576-586.

[RAH 03] RAHMAN M., WONG Y. S., ZAREENA A. R., 2003, "Machininability of titanium alloys", *JSME International Journal, Series C: Mechanical Systems, Machine Elements and Manufacturing*, 46(1), pp. 107-115.

[SAK 96] SAKURAI K., ADACHI K., KAMEKAWA T., OGAWA K., HANASAKI S., 1996, "Intermittently decelerated feed drilling of Ti-6Al-4V alloy", *Keikinzoku/Journal of Japan Institute of Light Metals*, 46(3), pp. 138-143.

[SAL 05] SALGADO M., LÓPEZ DE LACALLE L.N., LAMIKIZ A., MUÑOA M., SÁNCHEZ J.A., 2005, "Evaluation of the stiffness chain on the deflection of end-mills under cutting forces", *International Journal of Machine Tool and Manufacture,* 45: 727–739.

[SAL 31] SALOMON C., 1931, Process for the Machining of Metals of Similarly-Acting Materials when Being Worked by Cutting Tools, German patent No. 523594, 1931.

[SAX 71] SAXENA U. K., DEVRIES M. F., WU S. M., 1971, "Drill temperature distributions by numerical solutions", *Journal of Engineering for Industry*, 93(B4), pp. 1057-1065.

[SCH 92] SCHULZ H., MORIWAKI T., 1992, "High speed machining", *Annals of the CIRP*, vol. 41/2, pp. 637-643.

[SCH 95] SCHULZ H., HOCK S. T., 1995, "High speed milling of dies and molds - cutting conditions and technology", *Annals of the CIRP*, vol. 44/1, pp. 35-38.

[SCH 98] SCHULZ H., 1998, *High Speed Machining*, Société Française d'Éditions Techniques Sofetec, Paris.

[SHA 84] SHAW M. C., 1984, *Metal Cutting Principles*, Oxford University Press, New York.

[SIL 06] SILLER H., RODRIGUEZ C.A., AHUETT H., 2006, "Cycle time prediction in high-speed milling operations for sculptured surface finishing", *Journal of Materials Processing Technology*, vol. 174, no. 1-3, pp. 355-362.

[SMT 98] SMITH S., DVORAK D., 1998, "Tool path strategies for high speed milling aluminum workpieces with thin webs", *Mechatronics*, vol. 8, pp. 291-300.

[SRE 08] SREEJITH P.S., 2008, "Machining of 6061 aluminum alloy with MQL, dry and flooded lubricant conditions", *Materials Letters*, vol. 62, no. 2, pp. 276-278.

[STE 97] STEPHENSON D.A., AGAPIOU J.S., 1997, *Metal Cutting Theory and Practice*, Marcel-Dekker, New York.

[STE 06] STEPHENSON D.A., AGAPIOU J.S. 2006, *Metal Cutting Theory and Practice*, CRC Taylor & Francis, Boca Raton, FL.

[STR 04] STRENKOWSKI J. S., HSIEH C. C. and SHIH A. J., 2004, "An analytical finite element technique for predicting thrust force and torque in drilling", *International Journal of Machine Tools and Manufacture*, 44(12-13), pp. 1413-1421.

[SUH 73] SUH N. P., 1973, "The delamination theory of wear", *Wear*, vol. 25, pp. 111-124.

[SUH 86] SUH N. P., 1986, *Tribophysics*, New Jersey, Prentice Hall Inc.

[TLU 86] TLUSTY J., 1986, "Dynamics of High-Speed Milling", *ASME Journal of Engineering for Industry*, vol. 108, pp. 59-67.

[TÖN 86] TÖNSHOFF H.K., BUSSMANN W., STANSKE C., "Requirements on tools and machines when machining hard materials", *Proceedings of the 26th International Machine Tool and Research Conference*, 1986, pp.349-357.

[TRE 00] TRENT E. M., and WRIGHT P. K., 2000, *Metal Cutting*, Butterworth-Heinemann.

[VIG 97] VIGNEAU J., 1997, "Cutting materials for machining superalloys", *International CIRP/VDI Conference on Cutting Materials and Tooling*, Düsseldorf.

[VYA 97] VYAS A., SHAW M.C, 1997, "Chip formation when hard turning steel", *Manufacturing Science and Technology, ASME*, vol. 2, pp.21-28.

[WAN 94] WANG J-L., 1994, Development of New Drilling Force Models for Improving Drill Point Geometries, PhD Dissertation, University of Michigan.

[WAT 03] WATTS M. L., 2003, "High performance machining in aerospace", *Proceedings of the 2003, 4th International Conference on Metal Cutting and High speed Machining*, Darmstadt, Germany.

[WAT 77] WATANABE K., YOKOYAMA K., ICHIMIYA R., 1977, "Thermal analyses of the drilling process", *Bulletin of the Japan Society of Precision Engineering*, 11(2), pp. 71-77.

[WAT 85] WATSON A. R., 1985, "Drilling model for cutting lip and chisel edge and comparison of experimental and predicted results I-initial cutting lip model", *International Journal of Machine Tool Design and Research*, 25(4), pp. 347-365.

[WIL 96] WILKINSON R., 1996, *Aircraft Structures & Systems*, Longman, Edinburgh Gate, Harlow.

[YEN 04] YEN Y.C., JAIN A., ALTAN T., 2004, "A finite element analysis of orthogonal machining using different tool edge geometry", *Journal of Materials Processing Technology*, 146: 72-81.

[ZAR 01] ZAREENA A. R., RAHMAN M., WONG Y. S., 2001, "High speed machining of aerospace alloy Ti-6Al-4V", *International SAMPE Technical Conference*, 33, pp. 739-750.

Chapter 2

Analytical and Mechanistic Modeling of Machining Processes

2.1. The essential features of metal cutting processes

In metal cutting theory all operations performed with the wedge-shaped tool when a layer of metal is removed in the form of continuous or discontinuous chip can be conventionally divided into two general cases termed as orthogonal and oblique cutting.

2.1.1. *Orthogonal (2D) machining*

Typical practical cases of orthogonal machining include some machining operations performed with straight-edged tools such as planing (Figure 2.1a1) or surface broaching, turning of a narrow disk with radial feed (Figure 2.1a2) and turning of thin-walled tube with longitudinal feed. In all these cases the cutting edge is arranged to be perpendicular to the direction of relative tool-work motion and strain state is a 2D problem. Moreover, the tool cutting edge angle $\kappa r=900$ and tool inclination angle $\lambda s=0$. It is evident from the vector diagram of velocities shown in Figure 2.1b that by using the rule of continuity, the velocity of shearing v_s along the shear plane inclined

---

Chapter written by Wit GRZESIK.

at the shear angle Φ is the vectorial sum of the chip velocity v_{ch} (the sliding velocity at the tool-chip interface) given by equation [2.1] and the cutting velocity v_c. It can be determined from equation [2.2].

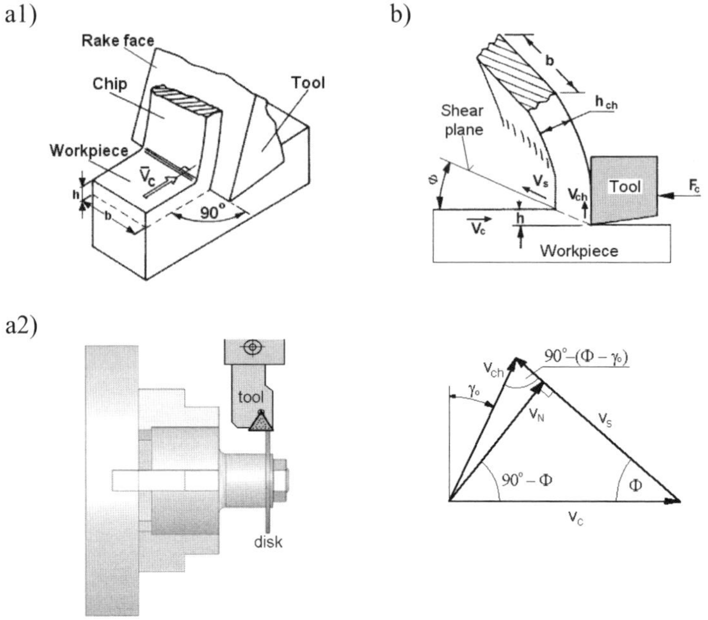

Figure 2.1. *Geometry of free orthogonal cutting processes with straight cutting edge (a) and velocity resolution (b) [GRZ08, GRZ98]*

$$v_{ch} = v_c \frac{\sin\Phi}{\cos(\Phi - \gamma_0)} \qquad [2.1]$$

$$v_s = v_c \frac{\cos\gamma_0}{\cos(\Phi - \gamma_0)} \qquad [2.2]$$

where Φ is the shear angle and γ_0 is the orthogonal rake angle.

2.1.2. *Oblique (3D) machining*

Figure 2.2a depicts a more general case of cutting known as oblique cutting, which is performed with tools having the tool cutting edge angle κr≠900 and tool inclination angle λs≠0. In contrast to orthogonal cutting, oblique cutting with tools having large inclination angle, typically for |λs|>300, is a 3D plastic deformation problem. Representative 3D cutting tools are a conventional lathe tool, a face-milling cutter and a twist drill with helical flutes. In a mechanistic model of oblique cutting (Figure 2.2b), the chip formation process consists of two parallel sub-processes including the plane plastic flow of the work material in the plane perpendicular to the cutting edge, similarly to orthogonal cutting (plane *Omn* in the *Olmn* coordinate system), and lateral plastic flow along the cutting edge (axis *l* parallel to the cutting edge *S*). The lateral material flow causes the chip to flow at the angle τ relative to the axis m1, i.e. it is deviated from the direction perpendicular to the inclined cutting edge.

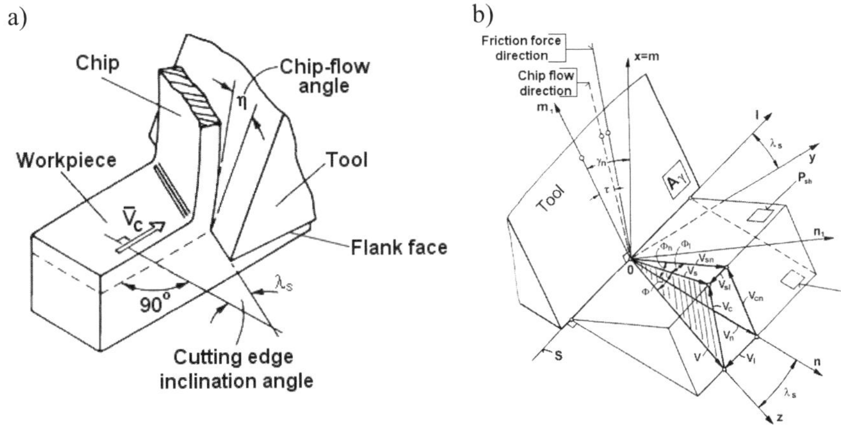

Figure 2.2. *Geometry of oblique cutting processes with straight cutting edge (a) and velocity resolution (b) [GRZ 08]*

As a result, the relationship between velocities, equivalent to equation [2.1], is [SHA 89]

$$v_{ch} = v_c \frac{\cos\lambda_s \sin\Phi_n}{\cos\tau\cos(\Phi_n - \gamma_n)} \qquad [2.3]$$

where Φn is the *normal* shear angle, γn is the normal rake angle and τ is the chip flow angle.

2.2. Plastic deformation and fracture

2.2.1. *Mechanisms of plastic deformation*

Plastic deformation in shearing, which is a general case of plastic behavior in metal cutting, can be realized as pure and plain (simple) shearing shown respectively in Figure 2.3. When the rotation angles of the deformed element ABCD (Figure 2.3a) are equal, i.e. $\gamma_{zx}=\gamma_{xz}$, such a deformation mechanism is called pure shearing. This is the case when uniform shortening and elongation take place along axes 3 and 1 respectively. On the other hand, plain shearing can be achieved only by rotating the opposite sides AD and BC of the square ABCD. As a result, the elementary square is transformed into a parralelogram and the circle inscribed into the square ABCD is converted into an ellipse (see the two sketches below in Figure 2.3b).

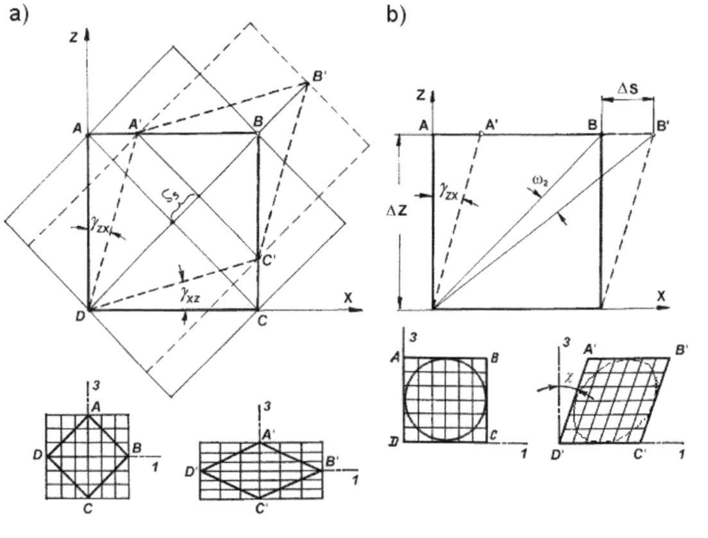

Figure 2.3. *Scheme of element deformation in pure (a) and plain (b) shearing*

2.2.2. *Measures of plastic deformation*

In the analysis of deformation state in the cutting zone, three different measures of plastic deformation can be of interest [SHA 89, GRZ 98]:

– shear strain simply defined, based on Figure 2.3b, as a tangent of rotation angle χ

$$\gamma_{sh} = tg\chi = \frac{\Delta s}{\Delta z} \quad [2.4]$$

where Δs is absolute deformation (linear displacement) and Δz is the small element dimension;

– normal (linear) strain which can be determined from Figure 2.3a as

$$\varepsilon = \frac{DB'-DB}{DB} = \left[\frac{1 + (1 + \gamma_{sh})^2}{2}\right]^{\frac{1}{2}} - 1 \quad [2.5]$$

– logarithmic (natural) strain recommended when normal strain is larger than 0.2, hence

$$\bar{\varepsilon} = \ln(1 + \varepsilon) = \frac{1}{2}\ln\left[\frac{1 + (1 + \gamma_{sh})^2}{2}\right] \quad [2.6]$$

2.2.3. *Material constitutive models*

Some models describing plastic behavior of ductile materials are selected in Figure 2.4. They were obtained by approximation of the real tensile/compression strain-stress curve.

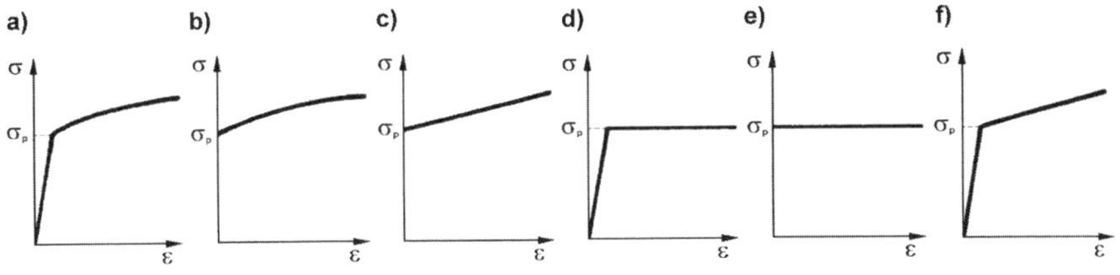

Figure 2.4. *Material models applied in plasticity theory of continuous medium: (a) exponential model, (b) rigid-plastic model with exponential strain-hardening, (c) rigid-plastic model with linear strain-hardening, (d) perfectly elastic-plastic model, (e) perfectly rigid-plastic model, (f) elastic-plastic model with linear strain- hardening [GRZ 98]*

In particular, if Figure 2.4a is replotted on log-log coordinates, a straight line is obtained having a slope *n* and thus beyond the yield point σ_p the Ludwik relation is valid, for which the following holds:

$$\sigma = \sigma_1 \varepsilon^n \quad [2.7]$$

where σ_1 is the stress corresponding to strain $\varepsilon=1$ and n is called the strain-hardening index. Equation [2.7] is widely used in analytical studies involving plastic flow in the cold working regime (see equation [2.11] by Oxley in Table 2.1).

2.2.4. *Mechanism of fracture in metal cutting*

All previously selected chip types can also be quantitatively related to types of material fracture based on the Fridman diagram, shown in Figure 2.5. This diagram utilizes the Tresca criterion for ductile fracture (maximum shear stress on the Y axis) and the maximum tensile stress (Saint Venant) criterion on the abscissa. Lines b, a and c adequately represent the brittle fracture strength (b), the yield point in shearing (a) and the yield point in shearing for perfectly ductile material (c). Additionally, line a' represents a specific case when the shear stress is artificially lowered by intensive cooling by supplying gaseous or liquid coolants. Inclined lines originating from such a coordinate system represent simple tension (1), simple compression (2) and simple torsion (3).

Material model/author(s)	Equivalent material flow stress
Johnson-Cook	$$\sigma_{eq} = \underbrace{\left(A + B\varepsilon_p^{\,n}\right)}_{I}\underbrace{\left(1 + C\ln\left(\frac{\dot{\varepsilon}_p}{\dot{\varepsilon}_p^0}\right)\right)}_{II}\underbrace{\left(1 - \left(\frac{T - T_r}{T_m - T_r}\right)^m\right)}_{III}$$ [2.8] where: I-elastic-plastic term; II-viscosity term; III-softening term ε_p – the equivalent strain, $\dot{\varepsilon}_p$ – the equivalent strain rate $\dot{\varepsilon}_p^0$ – the strain rate equal to 1.0 s^{-1}, T – the absolute temperature T_r – the room (ambient) temperature, T_m – the melting temperature of a specific metal or alloy, A, B ,C, n, m – five material constants.
Zerilli-Armstrong	I. For bcc (body-centered-cubic) metals: $$\sigma_{eq} = C_o + C_1\exp\left(-C_3T + C_4T\ln\left(\frac{\dot{\varepsilon}}{\dot{\varepsilon}_0}\right)\right) + C_5\varepsilon^n$$ [2.9] II. For fcc (faced-centered-cubic) metals: $$\sigma_{eq} = C_o + C_2\varepsilon^{0,5}\exp\left(-C_3T + C_4T\ln\left(\frac{\dot{\varepsilon}}{\dot{\varepsilon}_0}\right)\right)$$ [2.10] where: ε, $\dot{\varepsilon}$, $\dot{\varepsilon}_0$ are the same as in equation [2.8], C_0, C_1, C_2, C_3, C_4, C_5, n – seven material constants
Oxley	$$\sigma_{eq} = \sigma_1\varepsilon^n$$ [2.11] $$T_{mod} = \left(1 - \nu \cdot \log\frac{\dot{\varepsilon}}{\dot{\varepsilon}_0}\right)T$$ [2.12] where: σ_1 – the material flow stress at $\varepsilon = 1.0$ and n is the strain-hardening index. Values of σ_1 and ε depend on the velocity modified temperature T_{mod},, ν, $\dot{\varepsilon}_o$ – two material constants for a given material and range of testing conditions.
Marusich (Thirdwave AdvantEdge)	$$\bar{\sigma} = \sigma_f\left(\varepsilon^p\right) \cdot \left(1 + \frac{\dot{\varepsilon}^p}{\dot{\varepsilon}_o^p}\right)^{\frac{1}{m_1}}$$ [2.13] where $\bar{\sigma}$ is the effective von Mises stress, σ_f is the flow stress, ε^p is the accumulated plastic strain, $\dot{\varepsilon}_o^p$ is a reference plastic strain rate, m_1 is the strain rate sensitivity exponent. $$\sigma_f = \sigma_o \cdot \Theta(T) \cdot \left(1 + \frac{\varepsilon^p}{\varepsilon_o^p}\right)^{1/n}$$ [2.14] where: σ_o is the initial yield stress at the reference temperature T_o, ε_o^p is the reference plastic strain, *n* is the hardening exponent and $\Theta(T)$ is the thermal softening factor ranging from 1 at ambient to O at melt.

Table 2.1. *Mathematical expressions for material constitutive models [GRZ 08]*

Consequently, three types of material fracture can be distinguished, namely: brittle fracture in point A, partially ductile fracture in point C (because of the linear work-hardening effect (see τ_{max}-γ_{max} diagram on the right)) and perfect ductile fracture in point E. It can be assumed that these mechanical material states are equivalent to those responsible for the formation of discontinuous, segmented and continuous chips.

a) b)

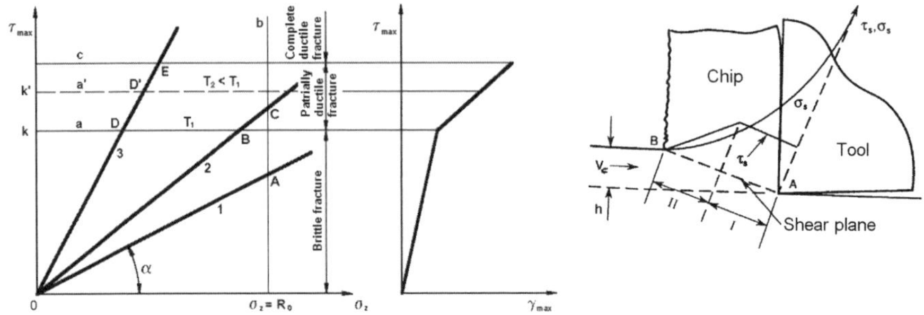

Figure 2.5. *(a) Fridman diagram of fracture mechanisms and (b) variation of stresses across the shear plane [GRZ 98, SHA 89]*

It was suggested by Shaw [SHA 89] that the real distribution of stresses on the shear plane is not uniform but their variations can be observed as in Figure 2.5b. In region I (nearest the cutting edge) τs is independent of normal stress on the shear plane (σs), while in region II, adjacent to the free surface of the chip, the ratio of $\tau s/\sigma s$ is approximately constant corresponding to Amontons' law of friction. These two regions should be connected by a transition region, where the density of microcracks increases to the point of intercommunication. Due to such a stress distribution, the mechanism of plastic flow involving the rewelding of microcracks would be expected to pertain to part I of the shear plane, where $\sigma s > \tau s$.

2.3. Chip formation in metal cutting operations

2.3.1. *Models for primary deformation zone*

In metal cutting, a chip formation area consisting of three characteristic zones I-III can be schematically presented as in Figure 2.6a. The extensive primary deformation (shear) zone (PDZ) I is distinguished within the OAB area by a number of slip-lines including the OFA line (the entry boundary) on which plastic deformation begins and the OCQB line (the exit boundary) on which chip material is entirely work-hardened. A material particle during deformation moves along the FQ path but the deformation process does not take place uniformly, rather it is more intensive between the shear band of Δx thick and the upper boundary OB. Then, the material adjacent to the tool-chip interface of OD length is subsequently deformed in depth Δ_1 due to intensive interfacial friction. The relevant OCD region is called the secondary deformation (shear) zone (SDZ) II. Additionally, the tertiary deformation zone (TDZ) III localized below the cutting edge in depth Δ_2 is separated.

a)

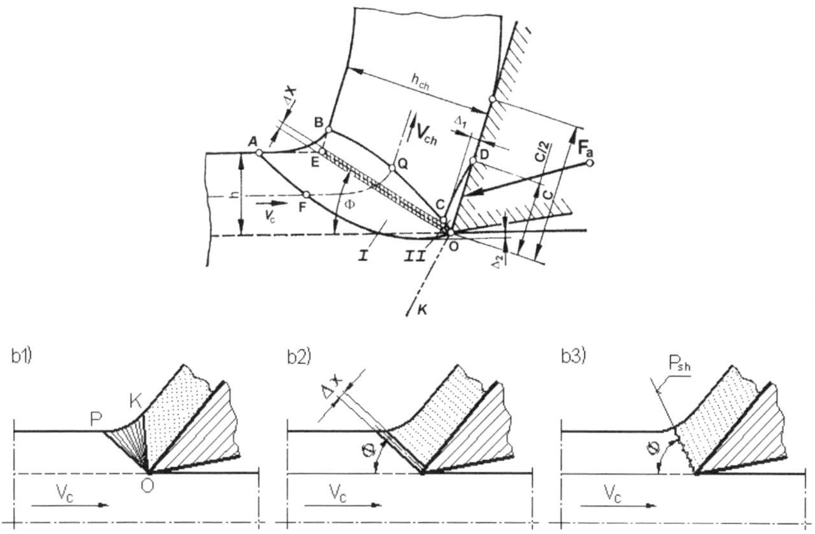

Figure 2.6. *(a) Schematic and (b1-b3) simplified models of the chip formation zone [GRZ 98]*

The engineering approach to the characterization of plastic deformation in the cutting zone is based on three basic simplifications (b1-b3 in Figure 2.6) including replacing curvilinear boundaries by straight lines OP and OK ("fan"-type or pie-shaped shear zone model proposed by Zorev [ZOR 66]), parallel-sided shear band inclined at angle Φ to the tool motion (model with parallel boundaries proposed by Oxley [OXL 89]) and one shear surface being a plane Psh (the shear plane model proposed by Merchant [SHA 89]). It should be noted that in the last case the work material is assumed to be perfectly plastic (incompressible), i.e. work hardening effect does not occur. Moreover, it ignores BUE, which may be present, as well as chip curl, and depicts tool face friction as elastic rather than plastic.

2.3.2. *Chip formation mechanisms*

2.3.2.1. *Continuous chip*

Continuous chips (item #1 in Figure 2.8) are common when most ductile materials, such as wrought iron, mild steel, copper and aluminum are machined at relatively high cutting speeds. The reason for this is that ductile metals and alloys do not fracture on the shear plane. It is noteworthy that cutting under these conditions is a steady-state process. The formation of a continuous chip takes place in the extended zone as illustrated in Figure 2.6. Continuous chip formation in ductile materials is associated with reduced cutting forces and generally results in good surface finish. In general, continuous chips may adopt many shapes – straight, tangled or with different types of helix. Often they have considerable strength due to intensive plastic deformation, and control of chip shape is one of the practical problems challenging machinists and tool designers. Usually, continuous chips with a built-up edge (BUE) are additionally distinguished at speeds at which the temperature at the chip-tool interface is relatively low and friction is so great that the chip material welds itself to the tool face. A BUE tends to grow until it reaches a critical size and then passes off with the underside of the chip and the new work piece surface.

According to International Standard ISO 3002/4, the chip compression ratio (Λ_h or k_h) is the ratio of the idealized chip thickness to the nominal thickness of the cut. It can also be determined using the weight method, as follows,

$$k_h = \frac{h_{ch}}{h} = \frac{m_{ch}}{\rho blh} \qquad [2.15]$$

where: m_{ch} is mass of the chip specimen, ρ is density of the work piece material, b, l and h are dimensions (width, length and thickness) of the uncut chip. In experimental work, the chip thickness can easily be measured with a ball-ended micrometer.

2.3.2.2. *Discontinuous chip*

Discontinuous chips (item #4 in Figure 2.8) are generally formed when machining brittle materials or materials containing points of stress concentration such as the graphite flakes in cast iron or the manganese sulfide inclusions in free-machining steels. They may also be produced when machining ductile materials at very low speeds and high feeds. Fracture occurs when the chip does not undergo plastic deformation and a crack propagates from the cutting edge (point *m*) to the free surface (point *n*). As a result, the chip breaks up into small segments in a cyclic manner.

2.3.2.3. *Segmented chip*

Segmented chips (item #2 in Figure 2.8) are characterized by regions of intense shear separated by regions of material with relatively lower deformation. Such chips are produced for most materials as cutting speeds are increased to high levels, but for materials with low thermal conductivity and low thermal capacity such as titanium alloys and austenitic stainless steels, segmentation takes place at more moderate speeds. In the latter case, cutting with a thermo-plastic instability, called adiabatic shear, occurs. As speed increases, the separation of the segments becomes more severe until complete separation of the segments (Figure 2.7b). Consequently, if the heat generated is not dissipated quickly enough, the catastrophic shear becomes localized and relatively less undeformed regions separate the narrow segments formed.

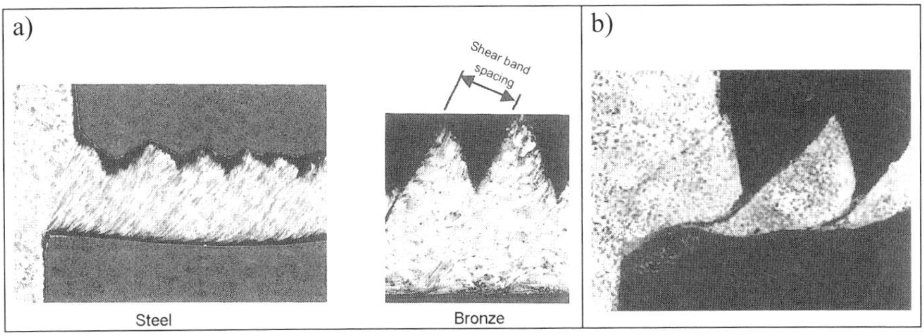

Figure 2.7. *Examples of differently segmented chips [BOO 06]: (a) partial segmentation, (b) complete segmentation*

There is a transition from one type of chip formation to another as speed is increased as for AISI 4340 steel with different hardness values. It can be observed that for more hardened material, the speed at which segmental chips are formed, and especially complete chip separation occurs as in Figure 2.7b, is lower.

2.3.3. *Chip formation models*

The formation of torn-off and shear-type chips can be explained in terms of material behavior during the deformation process using the appropriate stress-strain curves and relevant fracture mechanisms shown in Figures 2.5 and 2.8 respectively. In Figure 2.8 continuous chips are related to curve #1 with large plastic flow of the material and characteristic neck of a tensile specimen developed in region II. This phenomenon is termed by Shaw [SHA 89] as negative strain-hardening in the engineering stress-strain curve.

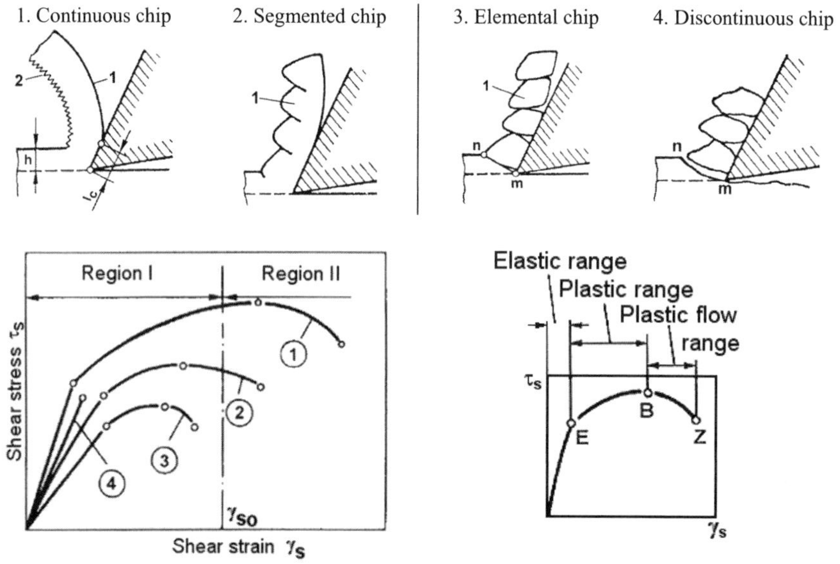

Figure 2.8. *Chip formation in terms of stress-strain curves; after Weber and Loladze [WEB 86]*

If the plastic flow is not so intensive and the shear strain slightly exceeds the critical value of strain γ_{so}, the onset of chip segmentation is observed (partially segmented chips vs. curve #2). When the entire shear strain decreases further, the chip segmentation starts to develop and the separation of chip segments begins (elemental chips vs. curve #3). In this case, the material necking takes place before fracture but at relatively lower shear stress. It should be noted that all three types of chips formed by shearing mechanisms could generally be related to the plastic deformation of the ductile work materials. In contrast, brittle materials only undergo elastic deformation (curve #4), which leads to producing discontinuous chips.

From the geometry of the cut (Figure 2.9), the shear angle is derived as a function of the chip compression ratio and the rake angle.

$$tg\Phi = \frac{\cos \gamma_o}{h_{ch} / h - \sin \gamma_o} = \frac{\cos \gamma_o}{k_h - \sin \gamma_o} \qquad [2.16]$$

Expressions for the shear strain can be obtained by assuming the deformation as a process of block slip, as shown in Figure 2.9b. Three different formulae for the shear strain are used:

$$\gamma_{sh} = ctg\Phi + tg(\Phi - \gamma_0) \qquad [2.17]$$

$$\gamma_{sh} = \frac{\cos\gamma_0}{\sin\Phi\sin(\Phi - \gamma_0)} \qquad [2.18]$$

$$\gamma_{sh} = \frac{k_h^2 - 2k_h\sin\gamma_0 + 1}{k_h \cos\gamma_0} \qquad [2.19]$$

It is interesting to note that the shear strain defined by equation [2.19] will be equal to zero only for k_h=1 and $\gamma o=90^0$. For example, for the rake angle $\gamma o=0^0$ the strain γ_{sh}=2, and covertly γ_{sh}=1 when the rake angle $\gamma o=36^0 52'$.

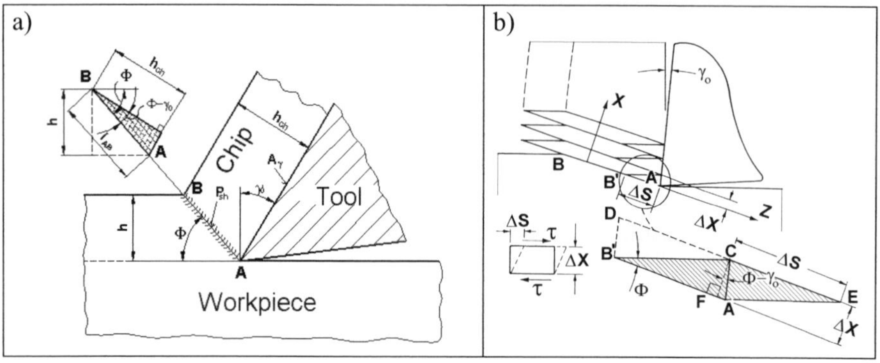

Figure 2.9. *(a) Geometry of the cut and (b) the shear-strain model [GRZ 98, TRE 00]*

The shear strain rate in cutting is defined as

$$\dot{\gamma}_{sh} = \frac{\gamma_{sh}}{\Delta t} = \frac{\Delta s}{\Delta x\, \Delta t} = \frac{v_s}{\Delta x} = \frac{\cos \gamma_0}{\cos(\Phi - \gamma_0)} \frac{v_c}{\Delta x} \qquad [2.20]$$

where: Δt is the time elapsed for the material element to travel a distance Δs along the shear plane, and Δx is the thickness of the shear zone (slip band).

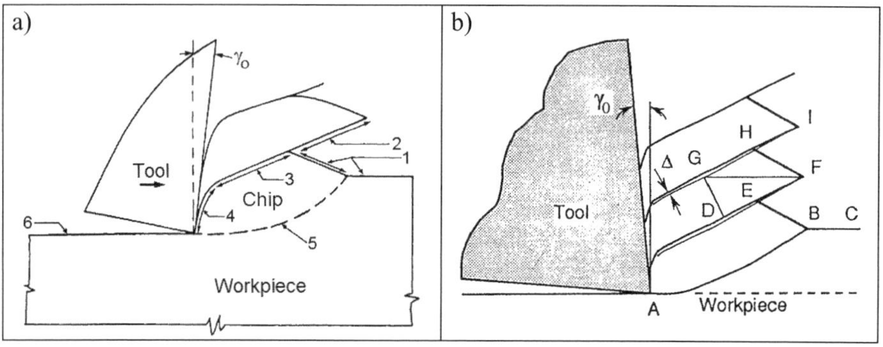

Figure 2.10. *(a) Schematic of the shear localized chip formation process [KOM 86, VYA 99] and (b) schematic diagram for calculating shear strains [TUR 82]. 1 – undeformed surface; 2 – part of the catastrophically shear failed surface separated from the following segment owing to the intense shear; 3 – intense shear band formed by catastrophic shear during the upsetting stage of the segment being formed; 4 – intensely sheared surface of a segment in contact with the tool and subsequently slid on the tool face; 5 – intense localized deformation in the primary shear zone*

The adiabatic shear theory suggests that the root cause of saw-tooth chip formation is a catastrophic thermo-plastic instability where the decrease in flow stress due to thermal softening, associated with an increase in strain, is more pronounced than the associated strain hardening. Figure 2.10a is a schematic diagram of the shear localized chip formation process proposed by Komanduri [KOM 86, VYA 99]. In general, two stages occurring simultaneously are involved in the formation of shear localized chips but the stress state and strain distribution are different in the two regions. This means that actually two shear planes, the auxiliary and primary shear planes, operate simultaneously [MAN 86]. One stage comprises plastic instability and strain localization in a narrow band (i.e. along line 5) in the primary deformation zone. The other stage involves gradual built-up of the segment with negligible deformation by upsetting the inclined work material by the advancing tool. During this stage intensive shear (even shear strain of about 60) between the segment just formed and the undergoing disturbance, takes place in a narrow band. A thermo-plastic instability occurring along dashed line (5) extends from the tool tip curving upward toward the free surface. As the tool advances, the material above line (5) is extruded into the shape (2,3,4,5), line (5) becoming surfaces (3) and (4). At the same time, inclined surface (1) becomes surface (2). In this model the final chip shape is completed after one cycle and then the process starts again. With an increase in cutting speed, the extent of contact between the segments decreases to the speed at which completely isolated segments of the chip will be formed. The chip formation by the catastrophic (adiabatic) shear was found to occur with Inconel 718 at a cutting speed above 100 m/min, hardened AISI 4340 steel at high speed of 260 m/min, Ti-6Al-4V titanium alloy at all speeds [KOM 86], case carburized steel of HRC 62 [VYA 99] and AISI 321 stainless steel at a cutting speed of 50 m/min [MAN 86]. It should be noted that the frequency of crack formation is very high in hard turning operations [VYA 99], hence a very rigid machining system is necessary in hard turning to accommodate shock occurring at the beginning and end of the cut.

Calculations of the shear strains occurring within the lamellae and along the shear fronts can be performed from geometric considerations of the shear strains illustrated in Figure 2.10b. The shear strain within the segments is termed the homogenous shear strain γ_h and the shear associated with the formation and propagation of shear bands is called the catastrophic shear strain γ_c. The maximum amount of homogenous shear strain applied in the first stage is given by

$$\gamma_h = \frac{GH}{GD} \quad [2.21]$$

and the amount of catastrophic (localized) shear is given by

$$\gamma_c = \frac{HI}{\Delta} \quad [2.22]$$

where Δ is the thickness of a thin shear band.

Thus, the total shear strain γ_{sb} in the shear bands is given by the sum of both components, i.e.

$$\gamma_{sb} = \gamma_h + \gamma_c \quad [2.23]$$

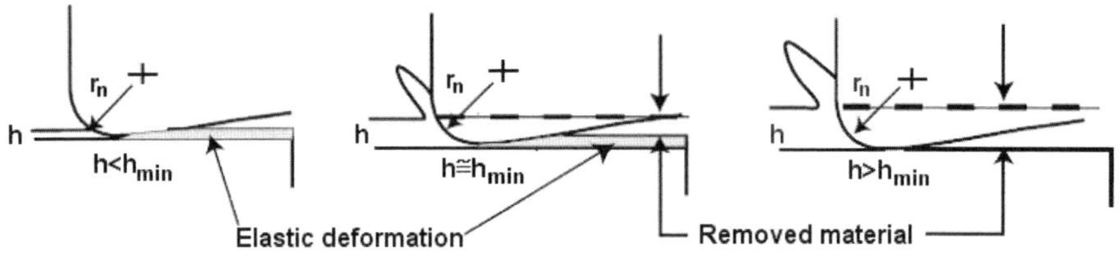

Figure 2.11. *Schematic of the effect of the minimum chip thickness on chip formation [CHA 06]*

Chip formation in micro-scale depends strongly on the minimum undeformed chip thickness (h_{min}) due to the fact that the chip may not be formed when the depth of the cut is less than h_{min}. As a result, simple scaling of the chip thickness variation model, assuming complete removal of the undeformed chip thickness (UCT) from the work piece and generating the chip, cannot be valid. In practice, the small UCT causes a large negative effective rake angle (γ_e) due to the small feed rate and the action of the rounded cutting edge, and, as a result, ploughing phenomenon occurs leading to the elastic recovery (also called spring back effect) or elastic-plastic deformation of the sub-surface layer. The transition from ploughing to chip formation is depicted in Figure 2.11.

Figure 2.12a shows a nanometric cutting model for the 2D molecular dynamics (MD) simulation of Cu and Al being cut by a rigid diamond cutting edge. The work material is divided into three different zones: the moving zone, the peripheral zone and the boundary zone. The forces produced by the Morse interaction potential determine the motion of the atoms in the moving zone. In addition, the dislocations generated near the cutting edge are supposed to move in a zigzag way, as shown in Figure 2.12b. The chip formation from one atomic layer is made possible by successive passing of the dislocations through the shear zone to the free surface (no dislocations remain in the chip). Figure 2.12c presents some results of the area-restricted molecular dynamics (ARMD) simulation with reduced computation time, applying to orthogonal machining of a (111) plane of a copper single crystal by a diamond tool with 3 nm edge radius, where the UCT and the cutting speed were set at 1 nm and 20 m/s respectively. Friction between the tool and work piece in the nanoscale machining was approximated by changing the cohesion energy of the interatomic potential function [MAE 95]. Figure 2.12c shows both the trajectories of the work atoms (left) and the instantaneous temperature of each atom at 60,000 time steps (equivalently after a cut distance of about 12 nm). It was found that, similar to macro-scale machining, all friction-dependent quantities such as the chip volume, the contact length, the number of atoms above 1000 K and the degree of the atomistic disturbance below the machined surface decrease with the reduction of the cohesion energy. This approach can also be applied to model tool wear in nanoscale machining. The ultimate surface roughness attainable in Cu and Al machining with the cutting edge radius of 5 nm, the UCT of 0.5 nm and the cutting speed of 200 m/s was equal to 0.5 and 0.8 nm, respectively [SHI 93].

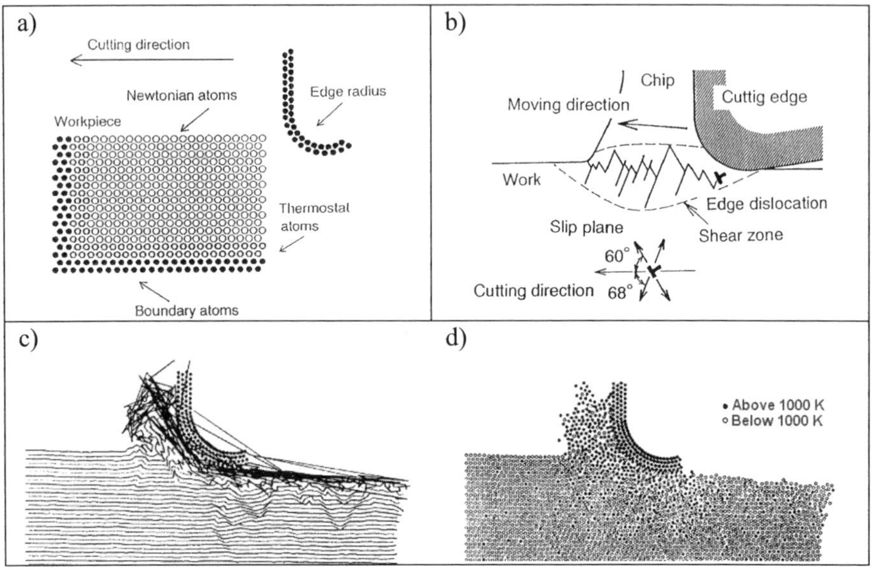

Figure 2.12. *Modeling of nanometric (micro) cutting using molecular dynamics simulation [SHI 93, MAE 95, CHA 06]. (a) Schematic model for molecular dynamics simulation, (b) schematic model of dislocation movement in shear zone, (c) behavior of work atoms and distribution of temperature for given potential shape*

2.3.4. *Relationships for shear angle*

The prediction of shear angle, which macroscopically quantifies the state of chip formation, is one of the pioneering solutions in metal cutting theory. The first model of the cutting process proposed by Ernst and Merchant, the shear in chip formation, was confined to the shear plane oriented by the movement of the tool. In shear angle theories many equations were proposed but it is still not clear which equation is most suitable, because of many factors that influence each other. Ernst and Merchant and Lee-Schaffer in their early studies of orthogonal machining, proposed to express the shear angle (Φ) as a function of the difference between the mean friction angle and the rake angle (Θ-γ_o) and ignore any effects of work hardening. The computation problem is that a unique function of (Θ-γ_o) for the shear angle does not exist. Merchant proposed to derive the shear angle by minimizing the cutting energy (the same is minimizing the cutting force) or maximizing the shear stress on the shear plane. The well-known equation represented by line 1 in Figure 2.13a results in:

$$\Phi = \frac{\pi}{4} - \frac{1}{2}(\Theta - \gamma_o) \qquad [2.24]$$

This optimum value of Φ gives $Fa=2khcot\Phi$, the normal pressure p on the shear plane being $kcot\Phi$.

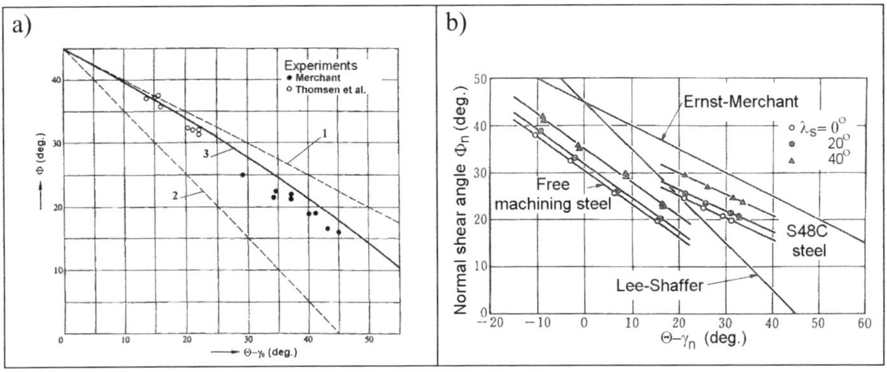

Figure 2.13. *Shear angle relations for orthogonal (a) and oblique (b) cutting [GRZ 89, USU 78]*

On the other hand, Lee and Schaffer proposed a simple slip-line field to describe the material flow in the triangular plastic zone ABC above the shear plane AB with slip-lines parallel and perpendicular to AB (Figure 2.14a). For force equilibrium of the free chip, it is required that the pressure on the primary shear plane is constant along the length of the shear plane and equal to *k*. Assuming (p/k)=1 and Δk=0, Lee and Shaffer's solution expressed by line 2 in Figure 2.13a is obtained as:

$$\Phi = \pi / 4 - (\Theta - \gamma_o) \qquad [2.25]$$

This means that AC is parallel to the line of action of the resultant cutting force, which is equal to $F_a=kh(1+cot\Phi)$. As shown in Figure 2.14a, neither equation [2.24] nor equation [2.25] is supported by the experiment. Childs *et al.* [CHI 00] suggest that in plotting the results as a graph of Φ against (Θ-γ_o), the process-centered view, i.e. (Θ-γ_o) is the complement of the angle between the shear plane and the tool rake face is a better choice. This is due to the fact that almost a single relation is observed between the friction coefficient and (Θ-γ_o).

Moreover, Figure 2.13b presents both the classical solutions for oblique cutting with the inclination angle ranging from 0 to 40^0. It is clear that the Lee-Schaffer's equation coincides better with the qualitative tendency in the experimental data obtained for both the carbon and free-machining sulfurized steels. However, similarly to the orthogonal cutting process, the quantitative agreement was not achieved.

According to Shaw [SHA 89], when the various shear angle equations are tested against experimental data, the points gathered for various rake angles lie along a straight line whose slope indicates that

$$\Phi = C - \Theta + \gamma_o \qquad [2.26]$$

The problem is the value of the constant C in equation [2.26], which is not always the same.

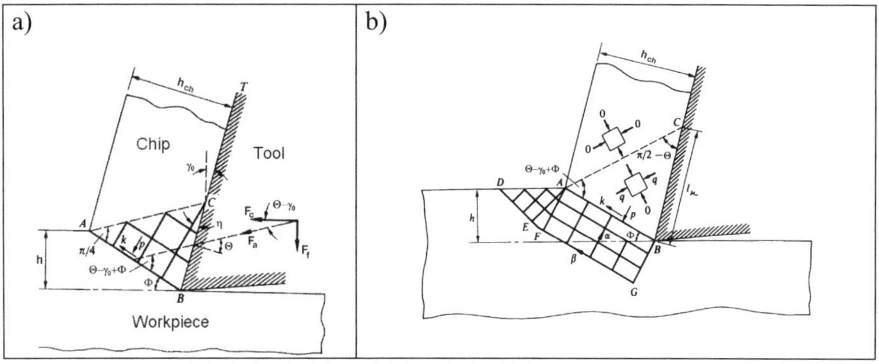

Figure 2.14. *(a) Slip-line fields for a rigid-perfectly plastic material proposed by Lee and Schaffer and (b) for linear work-hardening effect after Chakrabarty [CHA 87]*

The next step in predicting shear angle includes slip-line modeling with strain hardening effect and variation of the normal contact stress on the tool face. It should be noted that the straightness of the slip-lines results in a constant normal stress along the chip-tool contact and a sliding velocity equal everywhere to the chip velocity. An overview of slip-line fields for different material models and process variants, e.g. with restricted contact tools, chip breaking, the presence of work-hardening, changing of hydrostatic stress on the shear plane, stagnant zone, variable flow stress, friction factor, etc., is given by Childs *et al.* [CHI 00]. For example, Figure 2.15b illustrates the slip-line field obtained for assumption that the linear strain-hardening occurs when material is cut. For the shear angle Φ equal to or less than $\pi/4$, the slip-line field BADEFG consists of two regions with constant hydrostatic stresses, separated by a fan of angular span $\pi/4-\Phi$. In the triangular segment ADE, the slip-lines meet the stress-free surface AD at $\pi/4$, while in the region ABGF the slip-lines are parallel and perpendicular to AB. Consequently, the normal stress acting on the shear plane AB is given by

$$p / k = (1 + \pi / 2 - 2\Phi) \qquad [2.27]$$

The solution for the shear angle can be obtained by substituting the ratio of p/k in the general $p/k= \tan(\Phi\text{-}\Theta\text{-}\gamma_o)$ relationship. It is represented by straight line 3 in Figure 2.13a.

The third group of shear angle relationships was derived based on a simple upper bound model of orthogonal cutting [TRE 00]. Figure 2.15a shows the Rowe and Spick's model with two discontinuities, which accounts for the evidence of sticking friction and expresses friction at the tool face in terms of the relative shear stress (m). The general solution for the shear angle is obtained by differentiating the equation expressing the total work done according to the shear angle, as follows [TRE 00]:

$$m \sin^2 \Phi = \frac{1}{\chi} \cos \gamma_o \cos(2\Phi - \gamma_o) \qquad [2.28]$$

where ratio $m=\tau_c/\tau_s$ represents a friction condition like in strip rolling and χ characterizes contact length.

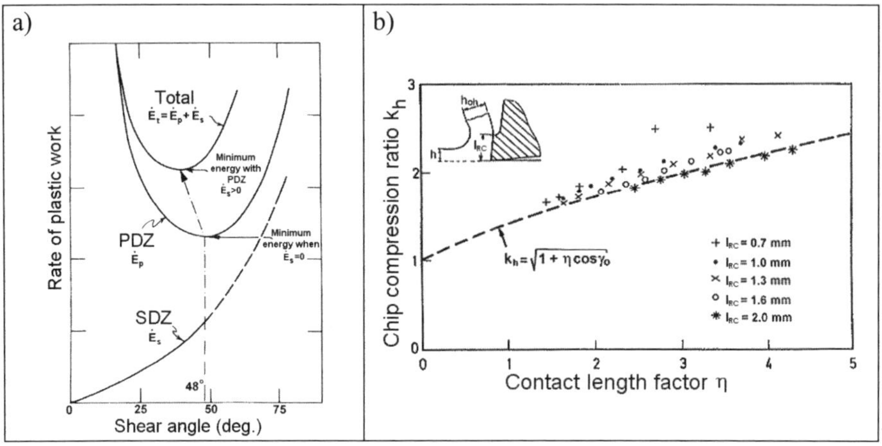

Figure 2.15. *(a)Shear angle and (b) chip compression ratio relationships using upper bound models [TRE 00, CHI 85]*

Equation [2.28] can be solved graphically for shear angle by plotting two families of curves corresponding to the left and right sides of the equation separately. Intersection of curves gives the value of shear plane angle. For example, for m=1 (the case of full sticking friction and contact length equal to UCT) the solution is identical to that proposed by Merchant [2.24].

In the new formulation proposed by Wright [TRE 00], the predicted shear angle utilizing the lowest upper bound to cause shear on severely work-hardened material of shear strength k_1 with defined power input, is given by

$$\cos(\Phi_o - \gamma_o)\sin\Phi_o = k_o / k_1\left[\cos(45^0 - \gamma_o/2)\sin(45^0 + \gamma_o)\right] \quad [2.29a]$$

where: shear angle Φ_o corresponds to the unstrained (softer) material with the shear stress of k_o.

De Chifree [CHI 85] has also used an upper bound approach, but instead of using a friction angle he introduced the contact length factor $\eta=l_c/h$ or l_{rc}/h for restricted contact tools (Figure 2.15b). Applying the minimum energy principle and assuming sticking friction conditions (m=1) in equation [2.28] the following relationship for the chip compression ratio can be derived

$$k_h = \sqrt{1 + \eta}\cos\gamma \quad [2.29b]$$

In addition, the corresponding shear angle relationship can be derived based on equation [2.29b].

When turning tests were run on an aluminum alloy with HSS tools ($\gamma_o=12^0$, $v_c=50$ m/min) the results shown in Figure 2.15b for various reduced contact lengths, are in excellent agreement with equation [2.29b].

$$p / 2k = \frac{k_h}{\cos \gamma_o} - \tan \gamma_o = \cot \Phi \qquad [2.30]$$

It should be noted that equation [2.30] is identical to the Merchant equation and this has proven that the simple shear plane model belongs to the types of upper-bound solutions.

2.3.5. *Chip flow rules*

The knowledge about chip flow direction across the rake face is necessary for determining the effective tool rake angle under more complex cutting arrangements, such as semi-orthogonal cutting with round-nosed tools and 3D oblique cutting, and primarily for planning the process in terms of efficient chip control. For ideal sharp-nose tools, the chip flows in the direction perpendicular to the main cutting edge. For the same type of tools but in free-oblique machining, involving 3D chip formation, the local chip flow direction follows Stabler's chip flow rule, that the chip flow angle measured from the normal to the cutting edge is equal to the tool inclination angle, i.e. $\eta_c=\lambda_s$ or $\eta_c=k\ \lambda_s$ where k is between 0.9 and 1 [KLU 79, BOO 06]. This means that Stabler's rule is invariable with the cutting speed and it neglects side flow by assuming the chip width is the same as the undeformed chip width. Better agreement with experimental data can be achieved by considering the normal rake angle ($\tan \eta_c=\tan \lambda_s \cos \gamma_n$) or the cutting speed ($\eta_c=\lambda_s\ v_c^{-0.08}$) as proposed by Russel and Brown [ARM 69] and Zorev [ZOR 66]. According to the rule of minimum cutting energy, in free-orthogonal cutting the chip should flow along the line with the maximum slope of the rake face (in the tool face orthogonal plane Pg), i.e. can be geometrically defined as

$$\tan \eta_c = \frac{\tan \gamma_f}{\tan \gamma_p} \qquad [2.31]$$

where γ_f is the tool side rake angle and γ_p is the tool back rake angle.

As a consequence, the effective rake can be calculated as

$$\sin \gamma_{ef} = \sin \eta_c \sin \lambda_s + \cos \eta_c \cos \lambda_s \sin \gamma_n \qquad [2.32a]$$

or after considering Stabler's rule

$$\sin \gamma_{ef} = \sin^2 \lambda_s + \cos^2 \lambda_s \sin \gamma_n \qquad [2.32b]$$

More complex equations which in addition to given cutting conditions require an extensive knowledge of factors which must either be predicted from machining theory or measured experimentally have also been proposed. For example, Armarego and Brown [ARM 69] using an equivalent shear plane model of the chip formation in oblique cutting showed geometrically that

$$\tan \eta_c = \frac{\tan \lambda_s \cos \gamma_n}{\tan(\Phi_n + \Theta_n)} + \tan \lambda_s \sin \gamma_n \qquad [2.33]$$

where Φ_n is the shear angle measured in the normal plane and Θ_n is the mean friction angle at the tool-chip interface measured in the same plane.

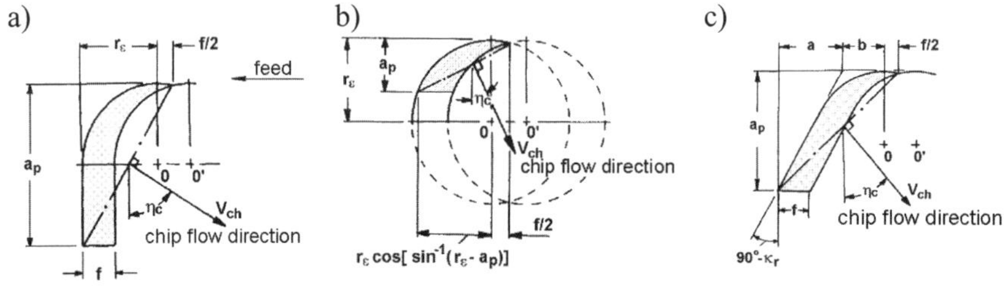

Figure 2.16. *Colwell's method of determining the chip flow angle for nose radius tools [OXL 89]*

For machining with nose radius tools, Colwell suggested a simplified geometrical method (Figure 2.16) which assumed that the direction of chip flow over the rake face would be normal to the chord joining the extreme points of the engaged cutting edge or to the major axis of the projected area of cut. However, the predicted results of chip flow direction satisfied only the cutting conditions with zero tool inclination and rake angles. The equation for the predicted chip flow angle obtained from this method can be expressed in three different ways depending on the cut geometry. For the case in which the tool cutting edge $\kappa_r=90^0$ and depth of cut a_p is greater than the tool nose radius r_ε (Figure 2.16a), the chip flow angle is

$$\tan \eta_c = \frac{a_p}{r_\varepsilon + f/2} \tag{2.34}$$

When the depth of cut is small enough in relation to the nose radius, so that only the curved edge is engaged (Figure 2.16b), then

$$\eta_c = \frac{a_p}{\left(2r_\varepsilon a_p - a_p^2\right)^{1/2} + f/2} \tag{2.35}$$

For a more general case, where the tool cutting edge $\kappa r\#90^0$ and the depth of cut is greater than the nose radius, as shown in Figure 2.16c, a more involved relationship is obtained, namely

$$\tan \eta_c = \frac{a_p}{a_p \tan \kappa_r + r_\varepsilon \tan(\pi/2 - \kappa_r/2) + f/2} \tag{2.36}$$

When using chip-grooved inserts, the chip groove geometry alters the chip flow from its natural state that would normally occur in machining with a flat-faced tool. In particular, under the same cutting conditions and tool geometry, the change in the chip-groove geometry will decisively result in different chip-flow patterns and therefore in different force distribution [EE 03]. Chip flow for a grooved tool occurs as a combination of chip side-flow and back-flow. Chip side-flow in a grooved tool is measured on the horizontal tool-face plane and is similar to that of a flat-faced tool, while the chip back-flow determines the degree of chip-streaming into the groove.

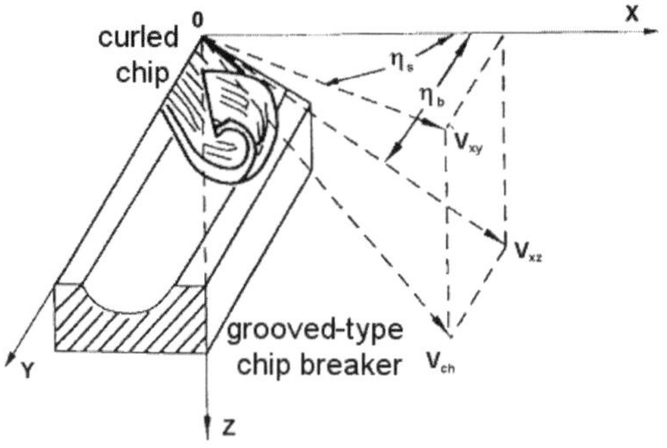

Figure 2.17. *3D chip flow with a grooved chip breaker [JAW 95]*

2.3.6. *Fracture in chip breaking*

Predicting 2D chip-breaking is based on the assumption that fracture of the chip modeled as a curved beam occurs when the tensile strain on the outer surface of the chip during bending will reach the ultimate tensile strain ε_{max} of the chip material. The relevant chip-breaking criterion, proposed by Nakayama [JAW 93, NAK 84] can be expressed as

$$\varepsilon_{max} \leq \frac{h_{ch}}{2}\left[\frac{1}{r_{ui}} - \frac{1}{r_{uf}}\right] = \varepsilon_b \qquad [2.37])$$

where h_{ch} is the chip thickness, ε_b is chip strain caused by bending, r_{ui} is radius of initial chip curl and r_{uf} is radius of final chip curl prior to fracture. This criterion is valid for chip-breaking occurring by chip/tool flank surface contact in an obstruction type chip-breaker.

Assuming, after Worthington *et al.* [JAW 93], as a first approximation, that for machining steels at high cutting speeds $h_{ch} \approx 2h$ (h is undeformed chip thickness) and the chip is effectively broken if it expands to twice its original radius, the chip breaking criterion can also be expressed by the equation

$$\varepsilon_{max} \approx \frac{h}{r_{ui}}\left(\frac{c-1}{c}\right) = \frac{h}{r_{ui}} A \qquad [2.38]$$

where for carbon steels and feeds ranging from 0.3 to 0.6 mm/rev, constant "*A*" varies from 0.166 to 0.5, i.e. a ratio of 1:3.

When fracture strain is known, it is possible to calculate the critical feed f_{cr} and the critical value of the depth of cut a_{cr} for the chip-breaking chart for groove tools with 2D chip-breaking features [BOO 06]. It should be noted that chip-breaking is very complex phenomenon depending strongly on the materials and metallurgical aspects of the cutting process [JAW 93], such as chip segmentation due to the adiabatic shear instability (the chip breaking is easier) and specific thermal effects caused by tool coatings with low thermal conductivity (due to a thermal bi-metallic effect, additional thermal stresses cause the chip to curl to a smaller radius).

2.4. Cutting forces and stresses

2.4.1. *Force distribution in the cutting zone*

In the general case, the force system exerted by a cutting tool is 3D, i.e. the resultant (total) cutting force F has three mutually perpendicular components shown in Figure 2.18a. Typically, the resultant cutting force (RCF) is resolved into geometrical and physical components. In the first case, the coordinate system is oriented based on the directions of primary and feed motions, i.e. the total force is resolved by perpendicular projection to these two directions. As a result, the cutting force and feed force denoted by symbols F_c and F_f are distinguished. Moreover, the back (passive) force F_p tending to push the tool away from the work in a radial direction and perpendicularly to the working plane is defined. This force is usually ignored for orthogonal cutting and the total cutting force is reduced to the active force F_a. As can be seen in Figure 2.18b, in orthogonal cutting the entire force system lies in a single plane and appropriate force components can easily be calculated by drawing the Merchant circle diagram. The force component in the direction of relative tool travel F_c determines the amount of work required to move the cutting tool a given distance. The feed force F_f determines the power required to feed drives installed onto a machine tool.

Geometrical resolution of the total cutting force is done by assuming that the chip is a body in stable mechanical equilibrium under the action of the forces exerted on it at the rake face and at the shear plane as shown in the right upper detail. The force components acting on the shear plane are F_{sh} and F_{shN}.

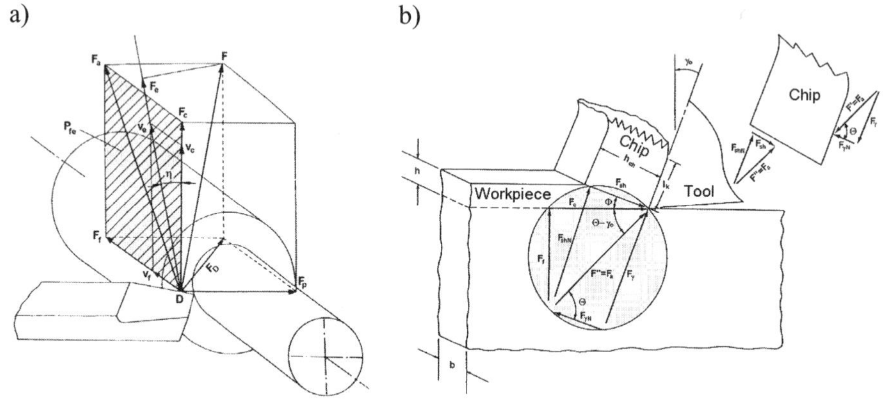

Figure 2.18. *Geometrical (a) and physical (b) force resolution in the cutting zone for orthogonal cutting with a continuous-type chip [GRZ 08]*

The shearing force F_{sh} represents the force required to shear the material on the shear plane. Force F_{shN} acts perpendicularly to the shear plane and results in an additional compressive stress being applied on the shear plane. They can be calculated from the following equations:

$$F_{sh} = F_c \frac{\cos(\Theta + \Phi - \gamma_0)}{\cos(\Theta - \gamma_0)} \qquad [2.39]$$

$$F_{shN} = F_c \frac{\sin(\Theta + \Phi - \gamma_0)}{\cos(\Theta - \gamma_0)} \qquad [2.40]$$

At the tool-chip interface, the force components F_γ and $F_{\gamma N}$ act parallel and perpendicularly to the face. F_γ, known as friction force (the tangential force at the rake), represents the frictional resistance met by the chip as it slides over the tool rake face. The friction force on the chip-tool interface:

$$F_\gamma = F_c \frac{\sin \Theta}{\cos(\Theta - \gamma_0)} \qquad [2.41]$$

where Θ is the friction angle.

The normal force on the chip-tool interface:

$$F_{\gamma N} = F_c \frac{\cos \Theta}{\cos(\Theta - \gamma_0)} \qquad [2.42]$$

2.4.2. *Models for cutting forces*

As depicted in Figure 2.19, the cutting forces are influenced by the cutting parameters as well as by tool geometry. Figure 2.19a shows a linear increase of all components of RCF due to relevant increase of the cross-sectional area of the uncut chip. On the other hand, force/feed curves are non-linear because increasing the feed results in decreasing the specific cutting pressure k_c [GRZ 08]. The kink in the force/cutting speed curves in the medium speed range illustrates the effect of a BUE. With steels, a BUE disappears when the speed is raised. The cutting forces are also influenced by tool geometry, the most important parameter being the rake angle. Increase in the rake angle lowers the force components but at the same time reduces the strength of the tool wedge and may lead to its fracture. Tools with high positive inclination angles generate high thrust forces causing the deflection of the work piece. In contrast, this force is very small when using tools with 90° cutting edge angle and machining of small-diameter work pieces is possible. The tool material and deposited coatings can also influence the cutting forces mainly by changes in the area of seized contact between the chip and the tool [TRE 00].

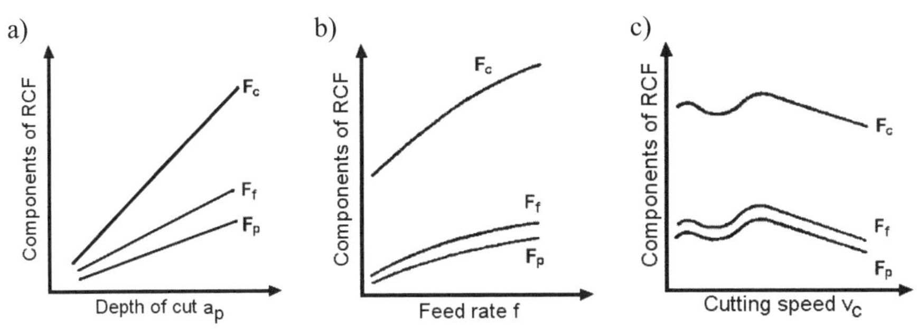

Figure 2.19. *Influence of cutting parameters on the components of resultant cutting force (RCF) [GRZ 08]*

The value of the cutting force can be predicted based on the specific cutting pressure $k_{c1.1}$ [KÖN 97] shear flow stress [BOO 06] or relationship between the works for cutting and tensile [ZOR 66], namely:

$$F_c = bh^{1-z}k_{c1.1} \qquad [2.43]$$

where: z is the slope of the line k_c=f(f) in log-log coordinates and $k_{c1.1}$ is the specific cutting pressure for the UCT=1 mm.

$$F_c = \frac{\tau_{sh} A_c}{\sin\Phi} \frac{\cos(\Theta - \gamma_{ne})}{\cos(\Phi + \Theta - \gamma_{ne})} \qquad [2.44]$$

where: A_c is cross-sectional area of the uncut chip, Θ is the friction angle and γ_{ne} is the working normal angle.

$$F_c = 0.6UTS(1 - 1.7Z)[(k_h - \sin\gamma_o)/\cos\gamma_o + \tan C] \qquad [2.45]$$

where: Z is percent reduction of area at fracture (necking) and average value of constant C =45^0.

2.4.3. *Ploughing force and minimum UCT*

It was experimentally proven that in cases such as cutting with a small undeformed chip thickness (UCT), cutting of materials with high elastic (Young's) modulus (and when tool wear at the flank progresses) and the forces acting on the flank-work piece contact should be considered in the total cutting force. It can be assumed that the force acting on the flank wear land VB_B (Figure 2.20) can be resolved into two components F_α and $F_{\alpha N}$ which are independent of the rake angle value. These forces can be determined using the method of extrapolation of F_c and F_f components on zero undeformed chip thickness (equation [2.46]) proposed by Zorev [ZOR 66].

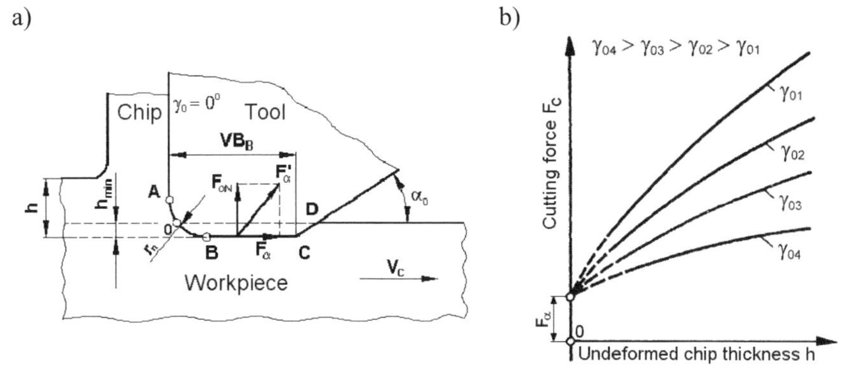

Figure 2.20. *(a) Force distribution on the work piece-flank contact and (b) projection of the cutting force on zero undeformed chip thickness [GRZ 08]*

As illustrated in Figure 2.20b, the Zorev method needs a few F_c/UCT curves to be obtained for different rake angles γ_{01}-γ_{04}.

From definition:

$$F_\alpha = F_{c(h\to 0)} \text{ and } F_{\alpha N} = F_{f(h\to 0)} \quad [2.46]$$

In addition, forces acting on the flank face can be calculated by solving the problem of penetration of a rigid punch into a plastically deformed half-space using the slip-line method [GRZ 98]. The normal force on the worn flank face:

$$F_{\alpha N} \approx c\sigma_p A_\alpha \quad [2.47]$$

where coefficient c=2.6-4, σ_p is the flow yield stress and A_α is the flank-work piece contact area. The tangential force on the flank face and

$$F_\alpha = \tau_{c\gamma} A_\alpha = (0.5 - 0.6)\text{UTS}\, A_\alpha \quad [2.48]$$

where $\tau_{c\gamma}$ is the main shear stress on the rake face and UTS is ultimate tensile strength.

The minimum undeformed chip thickness (h_{min}) was discussed in section 3.3 when considering the chip formation on a micro-scale. In general, it is assumed that h_{min} is proportional to the cutting edge radius and the coefficient of proportionality depends on the friction condition occuring, i.e.: $h_{min}=kr_n$. For dry friction k=0.1-0.2 and for semi-dry friction k=0.2-1.0. The transion from interfacial sliding to micro-cutting can be determined based on the molecular-mechanical theory of friction in the form of the Kragelsky-Drujanov equation [GRZ 98]

$$h_r \geq 0.5r_n\left(1 - \frac{2\tau_a}{Y_{sh}}\right) \qquad [2.49]$$

where h_r is the limiting depth of penetration of an indenter, r_n is the radius of indenter or rounded cutting edge, τ_a is the shear strength of the adhesive junction and Y_{sh} is the yield stress of strain-hardened bulk material (Y_{sh}=cY; Y is the yield stress of non-hardened material and c is the strain-hardening coefficient).

2.4.4. *Stresses on the shear plane*

Measurement of the forces and chip thickness make it possible to estimate the stresses on the shear plane for orthogonal cutting conditions, where a continuous chip is formed without built-up edge. With the assumption of uniform stress distribution and for known forces acting on the shear plane, the shear stress τ_s and normal stress σ_s are found to be

$$\tau_s = \frac{F_{sh}}{A_{sh}} = F_c \frac{\cos(\Theta + \Phi - \gamma_0)\sin\Phi}{bh\cos(\Theta - \gamma_0)} \qquad [2.50a]$$

$$\sigma_s = \frac{F_{shN}}{A_{sh}} = F_c \frac{\sin(\Theta + \Phi - \gamma_0)\sin\Phi}{bh\cos(\Theta - \gamma_0)} \qquad [2.50b]$$

where A_{sh} is the shear plane area (A_{sh}=bh/sinΦ), b is the width of cut and h is the uncut chip thickness. The ratio of the normal stress to the shear stress is

$$\frac{\sigma_s}{\tau_s} = tg(\Theta + \Phi - \gamma_0) \qquad [2.51]$$

The distribution of stresses along the shear plane is shown in Figure 2.21a. The shear stresses are uniformly distributed but the normal stresses are constant only for a zero or negative rake angle as indicated in case b. On the contrary, for positive rake angle (case a) normal stress decreases in the vicinity of the cutting edge and becomes tensile. Consequently, the stress ratio given by equation [2.51] is not constant.

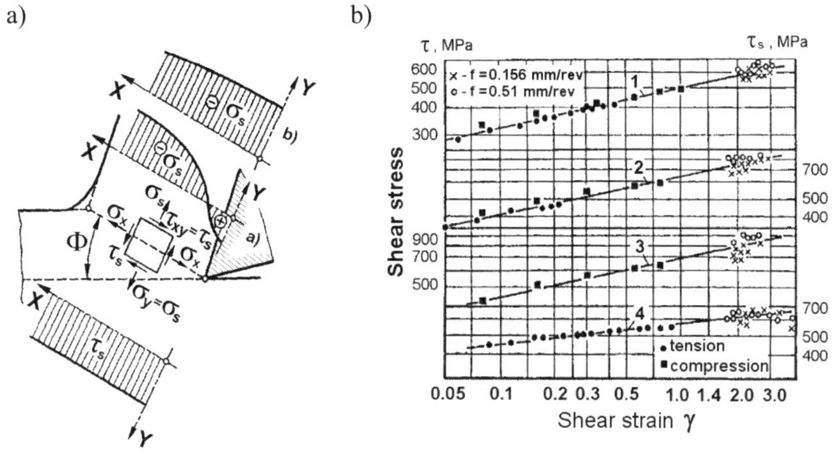

Figure 2.21. *(a) 2D model of stresses acting on the shear plane and (b) estimation of shear stress using simple tension and compression tests [GRZ 08, ZOR 66]*

For example, Figure 2.21b compares the shear stresses obtained for the cutting of different steels with stresses determined in both tensile and compression tests. Based on this comparison, Zorev [ZOR 66] concluded that the agreement exists only for very high shear strains, approximately in the range of γ_{sh}=2.5. Some other trials correlate the shear stress measured in cutting with the UTS or fracture stress.

2.5. Cutting energy and power

2.5.1. *Components of cutting energy*

The total work done by the cutting tool to remove metal, determined by the value of the force component Fc, is dissipated in the following forms for a sharp tool: (1) as work done in shear deformation, (2) as work done to overcome chip/tool interfacial friction and (3) as work done to generate a new surface in cutting, but the third form is often neglected. Practically all of the energy generated in metal cutting operations is dissipated in plastic deformation and friction.

In general, the energy required to perform primary and feed motions can be calculated using equations [2.52] and [2.53] respectively. As mentioned above, equation [2.54] defined two basic components of the cutting energy, which are required to perform plastic deformation (E_{sh}) and overcome friction (E_f) on the rake ($E_{f\gamma}$) and flank ($E_{f\alpha}$) faces.

$$E_c = \int_0^t F_c v_c dt \quad [2.52]$$

$$E_f = \int_0^t F_f v_f dt \quad [2.53]$$

$$E = E_p + E_f = E_{sh} + E_{f\gamma} + E_{f\alpha} \quad [2.54]$$

2.5.2. *Specific cutting energy/power*

Often, equivalent specific (volumetric) cutting energy (SCE) and its components are used as defined by equations [2.55] and [2.56], where V denotes the volume of the layer removed and A_c is cross-sectional area of the uncut chip.

$$e_c = \frac{E_c}{V} = \frac{F_c v_c}{bhv_c} = \frac{F_c}{A_c} = k_c \quad [2.55]$$

$$e_c = e_{sh} + e_{f\gamma} \quad [2.56]$$

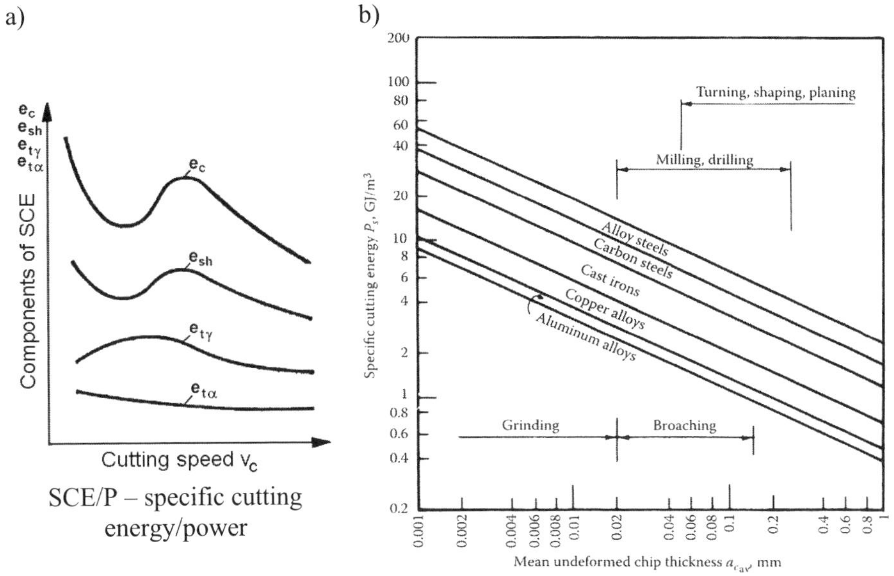

Figure 2.22. *(a) Influence of cutting speed on the components of specific cutting energy and (b) approximate values of the cutting energy for various materials and operations [BOO 06, GRZ 08]*

It is evident from equation [2.55] that the specific cutting energy is equal to the specific cutting pressure [AST 99]. On the other hand, components of specific cutting energy are equal

$$e_{sh} = F_c \frac{\cos(\Phi + \Theta - \gamma_o)\cos\gamma_o}{A_c \cos(\Phi - \gamma_o)\cos(\Theta - \gamma_o)} = \tau_{sh}\gamma_{sh} \quad [2.57]$$

$$e_{f\gamma} = F_c \frac{\sin\Theta}{A_c\, k_h \cos(\Theta - \gamma_0)} \quad [2.58]$$

where τ_{sh} is shear stress of the material in the shear plane, γ_{sh} is the shear strain and k_h is the chip thickness compression ratio.

Figure 2.22a illustrates changes of the components of SCE/P with cutting speed, and Figure 2.22b shows the influence of the undeformed chip thickness relative to typical machining operations and work materials. The shape of SCE/speed curve is the same as in Figure 2.19c. It is important to note that all finishing operations, especially performed on hard materials like grinding, are more energy consuming due to removing a very small layer of material.

2.6. Friction in metal cutting

2.6.1. *Models for tool/chip interface*

The tribological model of a cutting process can be schematically presented as in Figure 2.23. In this tribo-model with heavily loaded sliders, two/three characteristic regions can be distinguished [GRZ 98, TRE 00, BOO 06]:

1. The seized (sticking or plastic frictional contact) region A placed on the parts of both rake and flank faces adjacent to the tool cutting edge, in which the ratio A_r/A_a (real contact area to apparent contact area) approaches unity and continuous refreshing of the sliding surfaces takes place. In this region physically dry friction with chemically clean mating of active surfaces exists.

2. The sliding region (elastic frictional contact) region B extends from the end of the sticking region to the point where the chip loses contact with the tool (narrow region C). In contrast to the sticking region, in the region B dry technical sliding friction exists when supplying the cutting fluid boundary lubrication (the lubrication film is not completed) or fluid film lubrication. More accurate models consider the transition zone between A and B with mixed friction or lubrication conditions.

The existence of two different frictional contact modes was experimentally confirmed by Loladze [LOL 82] on the pictures of material plastic flow in the SDZ, as shown in Figure 2.23b. It is evident from Figure 2.23b that in the region adjacent to the cutting edge of l_{st} in length, the texture line 10 is parallel to the rake face and the velocity of a thin interfacial material layer v_{ch}^{st} approaches zero. This means that formation of the stagnation (dead) zone results from very strong adhesion between the fresh under-surface of the chip and rake face. The shear strain in this region reaches unusually high values of γ_{sh}=30-40. On the other hand, in the sliding region with uniform thickness Δ_2 of the SDZ, classical dry sliding friction occurs with the sliding velocity of v_{ch}^{sl}, which is proportionally lower than v_c.

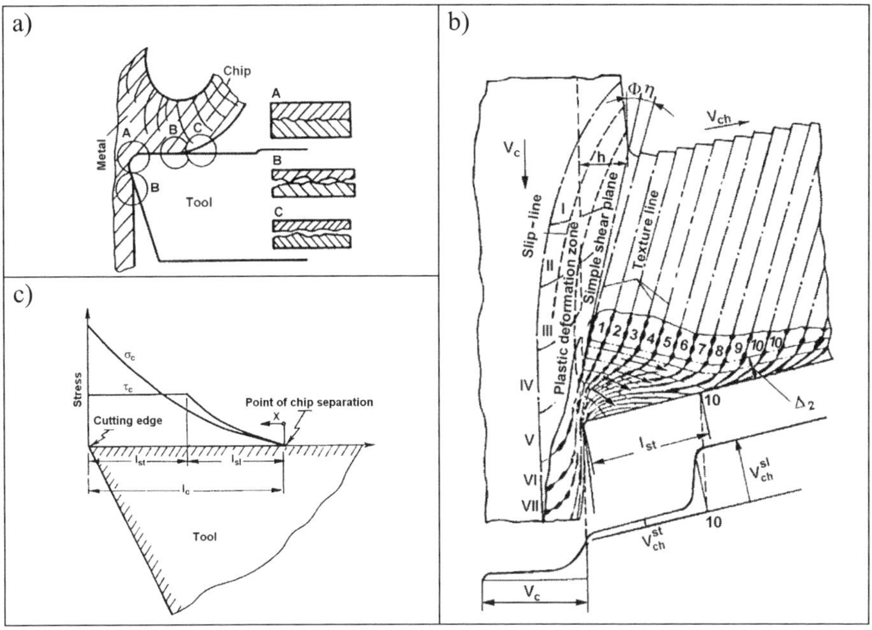

Figure 2.23. *Tribological characteristics of the cutting zone: (a) three zone model, (b) scheme of material plastic flow in the SDZ after Loladze, (c) simplified stress distribution along the tool-chip interface after Zorev [GRZ 98, LOL 82, ZOR 66]*

As shown in Figure 2.23c, in the absence of a built-up edge, sticking or seizure occurs at the interface near the tool edge and sliding takes place beyond the sticking region. For such a case, called Zorev's model, the normal stress distribution satisfies the power function $\sigma_c = qx^n$, where x is the distance from the point B where the chip breaks contact with the tool. Because the normal stress is distributed with the maximum at the cutting edge, falling to zero in the chip separation point, the expression for σ_c is

$$\sigma_c = \sigma_{c\,max}\left(x / l_c\right)^n \qquad [2.59]$$

where: $\sigma_{cmax} = \sigma_c$ for $x = l_c$.

The shear stress shows a lower maximum at the boundary between the sticking and sliding regions and a plateau across the sticking region. Due to this dualism, shear contact stress for the model from Figure 2.23c can be determined as follows:

$$\tau_c = \begin{cases} \mu\sigma_{c\,max}(x / l_c)^n \text{ for } 0 \le x \le l_{sl} \\ \mu\sigma_c \qquad\qquad \text{ for } l_{sl} \le x \le l_c \end{cases} \qquad [2.60]$$

where: l_c is tool-chip contact length, μ is coefficient of friction.

Usui and Shirakashi [CHI 00] proposed the friction model for the frictional shearing stress in the form of equation [2.61]:

$$\tau_f = k\left(1 - \exp\left[-\frac{\mu\sigma_n}{k}\right]\right) \qquad [2.61]$$

where: k is the shear flow stress of the local work material and the coefficient of friction μ is obtained from a series of experiments for different work piece-tool material pairs.

Furthermore, equation [2.61] can be modified by multiplying k with a friction factor m ($0<m<1$) and introducing an exponent n ($1<n<3$) to obtain a good saturation at k and mk for both non-free and free-cutting steels respectively, hence [CHI 00]:

$$\tau_f = mk\left(1 - \exp\left[-\left(\frac{\mu\sigma_n}{mk}\right)^n\right]\right)^{1/n} \qquad [2.62]$$

By using equations [2.61] and [2.62] it is possible to construct friction models with variable shear friction and variable friction coefficient at the entire tool-chip interface respectively [ÖZE 06].

2.6.2. *Tool/chip contact length*

It is obviously known that tool-chip contact length is one of the most important parameters controlling the cutting process and its estimation is based on several experimental, analytical and numerical methods such as experiments with split tools, slip-line theory and FEM simulation. Table 2.2 contains some important equations for calculation of the chip-tool contact length (in reality, the non-dimensional parameter l_c/h, termed as relative tool-chip contact length). This is based on the suggestion that the length of the sticking region can be assumed to be nearly equal to the chip thickness.

In general, the most important factor influencing l_c is the undeformed chip thickness (the feed) and this relationship is almost linear when machining carbon steels with cutting speeds of 150-300 m/min. The tool-chip contact length is also affected by the cutting speed (lower l_c values are recorded for higher cutting speeds) and depth of cut (higher l_c values are recorded for higher depths of cut). Contact length is also a function of both the mechanical properties of the work piece material and the tool material including coatings. Particularly, contact length decreases as the hardness of the work piece materials increases.

Author(s)	Equation for relative tool-chip contact length	
Lee and Schaffer [LEE 51]	$\frac{l_c}{h} = \frac{\sqrt{2}}{\sin\left[\tan^{-1}\left(\frac{\cos\gamma_o}{k_h - \sin\gamma_o}\right) \times \sin\left[45^0 + \tan^{-1}\left(\frac{\cos\gamma_o}{k_h - \sin\gamma_o}\right) - \gamma_o\right]\right]}$	[2.63]
Oxley [OXL 89]	$\frac{l_c}{h} = \frac{\sin\omega}{\cos\Theta\sin\Phi}\left\{1 + \frac{Cn}{3[1 + 2(1/4\pi - \Phi)] - Cn}\right\}$	[2.64]
Vinogradov [VIN 85]	$\frac{l_{st}}{h} = \frac{\sin\pi/4}{\sin\Phi_T\sin[\pi/4 + \Phi_T - \gamma_0]}$; $\frac{l_{st}}{l_c} = 0.4 - 0.56$	[2.65]
Abuladze [ZOR 67]	$\frac{l_c}{h} = \left(\frac{h_{ch}}{h}\right)^{0.1}\left[\frac{h_{ch}}{h}(1 - \tan\gamma_o) + \frac{1}{\cos\gamma_0}\right]$	[2.66]
Kachanov and Poletika [TOP 03]	$l_c = \frac{2h\cos(\gamma_o - \Phi)}{\sin\Phi}$ or $\frac{l_c}{h} = 2\left(\frac{h_{ch}}{h}\right)$ or $\frac{l_c}{h} = 2.05\left(\frac{h_{ch}}{h}\right) - 0.55$ (for $1 \le k_h \le 10$)	[2.67]
where: Φ is the shear angle, Φ_T is the shear angle calculated from mechanical properties of the work piece material, Θ is the friction angle, ω is the angle between the resultant cutting force and the shear plane, h is UCT, h_{ch} is the chip thickness, k_h is the chip thickness coefficient ($k_h = h_{ch}/h$); γ_o is the rake angle.		

Table 2.2. *Equations for calculation of the chip-tool contact length*

2.6.3. *Determination of friction coefficient*

In metal machining the frictional processes are very complex because of existence of very high normal loads. An increase in the normal load across the contacting surfaces produces an increase in the real contact area (A_r) until it is capable of supporting the applied load. Because a newly generated chip, having a highly chemically activated surface, slides over the rake face, adhesion (welding of asperities) is the predominant phenomenon at the interface and the frictional force stems from shear fracture of the bonded materials. The equivalent coefficient of friction is given by

$$\mu_c = \frac{F_t}{N_t} = \frac{\tau_t A_p}{\sigma_t A_p} = \frac{\overline{\tau}_c}{\overline{\sigma}_c} \qquad [2.68]$$

where: A_p is the apparent area of contact and τ_t and σ_t are the shear strength and yield pressure of the softer (chip) material.

In lightly loaded conditions (region of sliding contact) where Ap/Ar is very small, the Coulomb-Amonton laws of dry sliding friction hold. In case of relative motion between the highly loaded surfaces (region of sticking friction close to region of sliding friction) the frictional force is no longer linearly related to the normal force. At extremely high normal loads (region of sticking friction close to the cutting edge) the real contact can approach, or become equal to, the apparent area of contact. With these conditions, the coefficient of friction reaches a limiting value of 0.577 if it applies the von Mises plastic flow rule [BAI 75]. For Ap/Ar=1, seizure occurs, the frictional force is independent of the normal force and the coefficient of friction decreases with a further increase in normal load. In metal cutting, a simple method for calculating the mean normal $\overline{\sigma}_c$ and frictional stresses $\overline{\tau}_c$ from the shear plane model and the mean coefficient of friction (equation [2.68]), is used. By considering that for intensive friction $\sigma c \approx 2\tau sh(1.3 \pm \gamma 0)$, the maximum value of friction coefficient is given by:

$$\mu_{c\,max} = \frac{1}{2(1.3 \pm \gamma_0)} \qquad [2.69]$$

For example, for $\gamma o = \pm\pi/6$ values of $\mu cmax$ are equal to 0.64 and 0.27. The first value of 0.64 is close to the limiting value of 0.577 satisfying the von Mises plastic flow rule.

The total coefficient of friction on the rake face can also be estimated by adding the adhesion and deformation (mechanical) terms [GRZ 99]:

$$\mu_c = \mu_a + \mu_m \qquad [2.70]$$

2.6.4. *Relationships between friction and plastic deformation*

Metal cutting belongs to the range of mechanical processes in which plastic deformation is correlated with the intensity of friction in the contact zones. In general, process conditions influence these physical processes directly or indirectly. The first group includes such factors as the action angle $\omega=\Theta-\gamma_0$ (the angle between forces F_c and F_a in Figure 2.18b), the rake angle, cutting speed and mechanical properties of the work material. According to Zorev, such influences can be represented by combined temperature-strain-rate factor [ZOR 66]. Taking into account the minimum energy approach, the influence of angle ω can be expressed as follows:

$$2\Phi + \Theta - \gamma_o = 2\Phi + \omega = C \qquad [2.71]$$

where for the majority of carbon and alloy steels the constant $C=50^0\pm2^0$.

In order to correlate the shear angle Φ with the friction angle Θ, other formulae such as Merchant equation [2.24] and Lee-Schaffer equation [2.25] can also be applied as depicted in Figure 2.13.

2.7. Thermal modeling

2.7.1. *Heat partitioning*

Most thermal analyses of the cutting process are based on the moving heat source theory, for which it is necessary to know the fraction of the total cutting energy, conducted as heat to the tool, in order to estimate the interface temperature. In this theory, the chip is the body with a moving heat source and the heat partition coefficient (R_{ch}) defines the percentage of the heat entering the moving chip. Thus, the fraction ($1-R_{ch}$) provides the percentage of the dissipated energy going to the tool, i.e. the member that is stationary relative to the heat source. In general, three different methods of calculations of the heat partition coefficient are used, namely those proposed by Shaw: R_{SH} [SHA 89], Fujii and Kato: R_{KF} [KAT 99], and Reznikov: R_R [REZ 81].

$$R_{SH} = \frac{1}{1+\left[(0.754(\lambda_T / \lambda_W))/\left(A_a\sqrt{N_T}\right)\right]} \qquad [2.72]$$

In equation [2.72], thermal conductivities of work (λ_W) and tool (λ_T) materials, thermal number (N_T) and area shape factor (A) describe the thermal and geometrical features of the interface.

The thermal number is given by

$$N_T = v_{ch} l / 2\alpha_W \qquad [2.73a]$$

where v_{ch} is chip (sliding) velocity, $l=l_{nc}$ is natural contact length and α_W is the diffusivity of work (chip) material.

The shape factor (A) is determined as follows [SHA 89]:

– for estimating the average temperature, corresponding average value of this factor is equal to:

$$A_a = \frac{2}{\pi}\left[\sinh^{-1}\left(\frac{m}{l}\right) + \left(\frac{m}{l}\right)\sinh^{-1}\left(\frac{l}{m}\right) + \frac{1}{3}\left(\frac{m}{l}\right)^2 + \frac{1}{3}\left(\frac{l}{m}\right) - \frac{1}{3}\left\{\left(\frac{l}{m}\right) + \left(\frac{m}{l}\right)\right\}\left\{1 + \left(\frac{m}{l}\right)^2\right\}^{0.5}\right] \qquad [2.73b]$$

– the maximum value of factor A is

$$A_m = \frac{2}{\pi}\left[\sinh^{-1}\left(\frac{m}{l}\right) + \left(\frac{m}{l}\right)\sinh^{-1}\left(\frac{l}{m}\right)\right] \qquad [2.73c]$$

In equations [2.73b] and [2.73c], the width of the contact zone $m=a_p$ (a_p is the depth of cut) and the ratio m/l is the aspect ratio of the contact area. Kato and Fujii [KAT 99] originally defined heat partition for conventional surface grinding and for the thermal analysis in cutting a modified version is proposed [REZ 81]:

$$R_{KF} = \frac{1}{1 + \sqrt{(c_p \rho \lambda)_T / (c_p \rho \lambda)_W}} \qquad [2.74]$$

It should be noted that in equation [2.74], triple products, also known as *the heat transmission ratios*, for work (W) and tool (T) materials are introduced. The third formula uses both Peclet ($Pec=v_{ch}l_c/\alpha_{ch}$) and Fourier ($Fo=\alpha_w/(l_c v_{ch})$) number in order to consider the velocity and duration of frictional heat source [GRZ 03]. The relevant equation [2.75] includes equivalently thermal conductivities (λ_T and λ_W) and diffusivities (α_T and α_W) of both matching materials.

$$R_R = \frac{1}{1 + \left[(3\lambda_T / 2\lambda_W)\sqrt{\alpha_W / \alpha_T}\right]} \qquad [2.75]$$

Boothroyd (after Wiener) [BOO 06] proposed the proportion of heat conducted into the work piece β to be calculated using a unique function of R*tan*Φ (where Φ is the shear angle). The thermal number R is expressed as follows:

$$R = \frac{c_p \rho \cdot v_c \cdot h}{\lambda_W} \qquad [2.76]$$

where v_c is cutting speed, λ_W is thermal conductivity of work material and undeformed chip thickness h=f (f=feed rate).

Equation [2.76] is used to calculate temperatures in the primary deformation zone where heat is generated due to plastic deformation (shearing) required for chip formation. Theoretical calculation and experimentally determined values show that the coefficient β may be as high as 50% for very low rates of metal removal, materials with high conductivity like aluminum and copper alloys, and small shear plane angles. However, for high MRRs and machining steel it is about 10-15%.

2.7.2. *General analytical model for cutting temperature*

In general, the calculating concept is based on the well-known principle of the simultaneous action of two independent heat sources, which suggests that the total heat flux is generated by aggregation of the plastic deformation and sliding friction effects. Thus, the average interface temperature is defined as the sum of the mean shear-plane temperature ($\overline{\Theta}_s$) and the mean temperature rise due to friction ($\Delta\overline{\Theta}_f$), i.e.:

$$\overline{\Theta}_t = \overline{\Theta}_s + \Delta\overline{\Theta}_f \qquad [2.77]$$

As an analogy to equation [2.77], the maximum interface temperature is the effect of proper temperature peaks, i.e.:

$$\Theta_{max} = \Theta_{s\,max} + \Theta_{f\,max} \qquad [2.78]$$

2.7.3. *Temperature in the primary and secondary deformation zone*

The temperature rise for the primary zone is calculated based on the rate of working in a discrete parallel sided primary zone using an upper bound type analysis [TRE 00].

$$dW / dt = \tau_p A_p v_s \qquad [2.79]$$

where: τ_p is the average shear strength in the primary zone, A_p is the area of the shear plane and v_s is the shear velocity (see Figure 2.1b and equation [2.2]).

The calculation scheme used consists of three steps. Firstly, the thermal number R is calculated using Boothroyd's formula [2.76]. The next two steps of calculations are based on the theory of similarity elaborated for metal cutting purpose by Silin [GRZ 04]. This enables the estimation of the maximum and mean temperatures occurring at the shear plane. It should be noted that the theory of similarity enables us to calculate this temperature more accurately than with Oxley's theory because more calculation formulae are proposed depending on the value of the product *R*tan*Φ*. Maximum temperature at the shear plane can be calculated as:

$$\Theta_{s\ max} = \frac{\tau_{sh}}{c_p \rho \tan \Phi} \operatorname{erf} \sqrt{\frac{R \tan \Phi}{4}} \qquad [2.80]$$

where: τ_{sh} is shear flow stress, Φ is shear angle.

Then, the mean temperature at the shear plane is equal to:

$$\overline{\Theta}_s = 0.685 (R \tan \Phi)^{0.07} \Theta_{s\max} \quad \text{for } R \tan\Phi \le 5 \qquad [2.81a]$$

$$\overline{\Theta}_s = 0.620 (R \tan \Phi)^{0.13} \Theta_{s\max} \quad \text{for } 5 \le R \tan\Phi \le 20 \qquad [2.81b]$$

$$\overline{\Theta}_s = 0.820 (R \tan \Phi)^{0.04} \Theta_{s\max} \quad \text{for } R \tan\Phi \ge 20 \qquad [2.81c]$$

Alternatively, expressing equation [2.80] in terms of measurable cutting quantities is

$$\Delta\Theta_p = \frac{(F_c \cos\Phi - F_f \sin\Phi)\cos\gamma_0}{c_p\rho b_c h_c \cos(\Phi - \gamma_0)} \qquad [2.82]$$

where F_c and F_f are cutting forces, Φ is the shear angle, γ_0 is the rake angle and $b_c h_c$ is the product of the undeformed chip width and thickness.

Many methods have been used to calculate temperatures and temperature gradients at the tool rake/chip interface. Wright *et al.* [TRE 00] have derived the equation for the temperature as a function of distance along the tool-chip interface in the region of seizure.

$$\Theta(x,0) = \frac{2\tau_s\dot{\gamma}_s h}{\rho c}\left(\frac{x}{\pi\alpha v_c}\right)^{\frac{1}{2}} \qquad [2.83]$$

where τs is the secondary shear stress, $\dot{\gamma}_s$ is the strain rate in secondary shear, h is the thickness of secondary shear layer and α is the thermal diffusivity.

According to Shaw [SHA 89] the temperature increment due to the action of the frictional heat source (l_c is the tool-chip contact length) can be determined as

$$\Delta\overline{\Theta}_f = \frac{0.377 R q l_c}{\lambda_W\sqrt{N_T}} \qquad [2.84]$$

2.8. Mechanistic and predictive models for orthogonal and oblique cutting

2.8.1. *Mechanistic models*

Mechanistics models of machining processes are aimed at the accurate prediction of dynamic cutting forces for further estimating other quantities of the cutting process including tool life, cutter and part deflection, surface quality and process stability. These models can be incorporated into a feedback loop involving the dynamics of the machine structure for the analysis of machine tool stability. The main assumption in the mechanistic approach is that the cutting forces are proportional to the uncut chip area and the constant of proportionality (cutting pressure) depends on the cutting conditions, tool geometry and work material properties. The magnitudes of the force components, normal and parallel to the rake face of the tool, are defined by [BOO 06, STE 97]:

$$F_n = K_n A_c \text{ and } F_f = K_f A_c \qquad [2.85]$$

where K_n is specific normal pressure, K_f is specific friction pressure and $A_c = f \times a_p$ is the uncut chip area (UCA).

It should be noted that the resulting cutting force is perpendicular to the rake face and the friction force lies in the plane of the rake face in the direction of chip flow defined by the chip flow angle τ_γ (see Figure 2.1). In terms of these parameters the tangential (F_c), axial (F_f) and radial (F_p) forces are given by:

$$F_c = K_n A_c \left[\cos\gamma_p \cos\gamma_n + K_f\left(\cos\tau_\gamma \sin\gamma_n + \sin\tau_\gamma \sin\gamma_p\right)\right] \qquad [2.86a]$$

$$F_f = K_n A_c \left[-\cos\gamma_p \sin\gamma_n + K_f\left(\cos\tau_\gamma \cos\gamma_n\right)\right] \qquad [2.86b]$$

$$F_p = K_n A_c \left[-\sin\gamma_p + K_f\left(\sin\tau_\gamma \cos\gamma_p\right)\right] \qquad [2.86c]$$

In this set of equations, γ_p is the back rake angle and γ_n is the normal rake angle.

This overall modeling approach has been adapted to a wide range of 3D cutting processes such as turning, boring, milling and drilling. In quasi-static analyses of face milling, the variation of the UCA with the cutter rotation is approximated by a sinusoidal function and forces are estimated by multiplying the UCA by measured cutting pressure. Forces are calculated for each insert embedded in the cutter as a function of the cutter rotation angle φ and total forces F_x, F_y and F_z are obtained by adding together the forces exerted on the individual inserts which are cutting at a particular time. The tangential, radial and axial forces acting on an insert *i* when the cutter rotation angle is φ are given by [STE 97]:

$$\begin{Bmatrix} F_t(i,\varphi) \\ F_r(i,\varphi) \\ F_a(i,\varphi) \end{Bmatrix} = K_t C_1(i,\varphi) a_e(i,\varphi) \begin{bmatrix} 1 + K_r \dfrac{\cos\tau_\gamma \tan\gamma_p}{\cos\gamma_f} \\ -\tan\gamma_p + K_r \dfrac{\cos\tau_\gamma}{\cos\gamma_f} \\ -\dfrac{\tan\gamma_f}{\cos\gamma_p} + K_r \dfrac{\sin\tau_\gamma}{\cos\gamma_f \cos\gamma_p} \end{bmatrix} \qquad [2.87]$$

where K_t and K_r are empirical cutting pressures in the tangential and radial directions, τ_γ is the chip flow angle, γ_f and γ_p are the axial and radial rake angles of the cutter, $a_e(i, \varphi)$ is the effective axial depth of cut and $C_1(i, \varphi)$ is the UCT for insert *i* at a rotation angle of φ.

In quasi-static modeling of end milling forces, the cutting edge of each flute of the cutter is divided into N axial elements of elemental height Δz and the tangential and radial forces on cutter at time t, $F_t(t)$ and $F_r(t)$ are given by [STE 97]:

$$F_t(t) = \sum_{i=1}^{N}\left[\sum_{j=1}^{n_t}\left(K_t t_{cij}(t)\right)\Delta z\right] \quad [2.88a]$$

$$F_r(t) = K_r F_t(t) \quad [2.88b]$$

where n_t is the number of flutes on the cutter, K_t and K_r are empirical cutting pressures and $t_{cij}(t)$ is the UCT for flute *j* at element *i* at time t.

In the case of drills with helical flutes, all variables included into equations for the force components (the cutting speed, the UCT and the rake and inclination angles) are functions of the radial distance from the drill axis to the center of the *i*th segment, r_i.

2.8.2. *Strain rate analysis*

A number of analyses of the metal cutting process using a thick shear-zone model have been developed by application of slip-line field modeling (see section 2.3.4) and the effects of flow stress varying with strain (strain hardening) and, in many cases, with strain rate and temperature. Pioneering work in this area, in which the high strain-rate/high temperature flow stress and thermal properties of the work material are applied, was done by Oxley [OXL 89]. Assuming that in the primary deformation zone variations of hydrostatic stress are dominated by flow stress variations, the direction of the resultant force Fa (see Figure 2.18b) across the shear plane, near the center of parallel-sided zone, is given by

$$\tan(\Phi + \Theta - \gamma_0) = 1 + 2(\pi / 4 - \Phi) - Cb \quad [2.89]$$

where C is the constant in an empirical strain-rate relationship

$$\dot{\gamma}_{max} = Cv_s / l_s \quad [2.90]$$

in which v_s is the shear velocity and l_s is the length of the shear plane and *b* is the strain-hardening index of the material in Ludwik equation [2.7].

Moreover, strain rate and temperature are combined into a single function, known as the velocity modified temperature T_{MOD} (K):

$$T_{MOD} = T\left(1 - \nu \log\left(\dot{\varepsilon} / \dot{\varepsilon}_0\right)\right) \quad [2.91]$$

where a material constant ν is about 0.09 and a reference strain rate $\dot{\varepsilon}_0$ is equal to 1. In this approach, a set of equations for the shear angle, the temperature at the shear plane, the strain at the shear plane, the average temperature at the tool/chip interface and the tool-chip contact length [OXL 89, BOO 06] is sufficient to calculate cutting forces, temperatures, etc. for given cutting conditions as long as the appropriate work material properties are known.

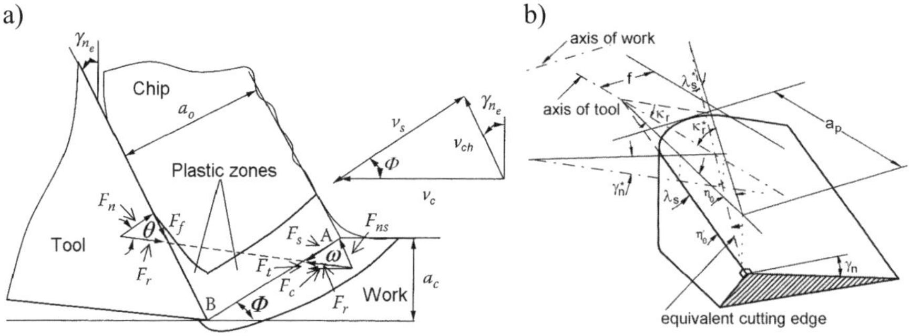

Figure 2.24. *(a) Predictive model for orthogonal cutting and (b) equivalent cutting edge with appropriate tool angles [OXL 89, OXL 98]*

$$\left.\begin{aligned} \lambda_s^* &= \sin^{-1}\left(\cos \eta_0' \sin \lambda_s - \sin \eta_0' \sin \gamma_n \cos \lambda_s\right), \\ \gamma_n^* &= \sin^{-1}\left(\frac{\sec \eta_0' \sin \lambda_s - \sin \lambda_s^*}{\tan \eta_0 \cos \lambda_s^*}\right), \\ \kappa_r^* &= \kappa_r + \eta_0 \end{aligned}\right\} \quad [2.92]$$

It is of practical interest that the above-mentioned orthogonal theory can be extended to predict the chip flow direction and cutting forces in oblique machining with nose radius tools by using the concept of an equivalent cutting edge shown in Figure 2.24 [OXL 98]. As a result, the equations [2.92] for the equivalent normal rake γ'n, inclination angle λ's and side cutting edge angle κ'r are obtained. The tool angles associated with the equivalent cutting edge, together with the predicted values of F_c and $F_t(F_f)$ and the value of chip flow angle $\eta_c^* = \lambda_s^*$, are then used to determine the force $F_r(F_p)$ which is normal to F_c and F_t.

$$F_r = \frac{F_c\left(\sin\lambda_s^* - \cos\lambda_s^* \sin\gamma_n^* \tan\eta_c^*\right) - F_t \cos\gamma_n^* \tan\eta_c^*}{\sin\lambda_s^* \sin\gamma_n^* \tan\eta_c^* + \cos\lambda_s^*} \quad [2.93]$$

$$L = AT_f^{-B} \quad [2.94]$$

where L is the tool life, T_f is the tool flank temperature and A and B are constants.

In addition, the predictions can be extended to the tool life (L) based on the data of Takeyama and Murat [GRZ 08] in the form of equation [2.94], the cutting conditions which give plastic deformation of the cutting edge, surface finish and, in general, to determing optimum cutting conditions based on objective criteria such as minimum cost, maximum production rate, etc.

Armarego [ARM 00] developed the concept of predictive modeling of various technological performance measures for the wide spectrum of machining operations used in practice (ball-end milling, tapping, drilling, form turning operations). This approach involves the development of generalized mechanics for analyses of the cutting processes for machining with single edge and multi-edge (form) tools and the establishment of a generic database of basic cutting quantities and edge-force coefficient. The models for turning, drilling and milling operations, as well as machining with form and rotary tools, are developed.

2.8.3. *Similarity method*

The concept of similarity method for use in metal cutting is originated from Silin [GRZ 08] and popularized by Astakhov [AST 00]. It is based on a set of similarity numbers including:

1. The A-criterion, which characterizes the heat quote of the chip relative to the entire amount of heat generated in the deformation zone.

$$A = \frac{h_1 b_1 (c\rho)_w \theta_c}{F_c} \qquad [2.95]$$

where h_1 and b_1 are the uncut chip thickness and its width respectively, $(c\rho)_w$ is the volume-specific heat of work piece material J/(m^3°C), θ_c is the cutting temperature °C and F_c is the tangential cutting force N.

2. The B-criterion is represented by:

$$B = \tan \Phi \qquad [2.96]$$

where Φ is the shear angle.

3. The Peclet number calculated from:

$$Pe = \frac{v_c h_1}{\alpha_w} \qquad [2.97]$$

where $v_c h_1$ characterizes the tool-chip contact conditions and α_w is the diffusivity of work piece material (m^2/s).

4. The D-criterion which characterizes the uncut chip area:

$$D = \frac{h_1}{b_1} \qquad [2.98]$$

5. The E-criterion which gives the relationship between the cutting edge radius r_n and the UCT:

$$D = \frac{r_n}{h_1} \qquad [2.99]$$

6. The F-criterion which characterizes the influence of tool geometry by considering the thermal properties of tool and work materials:

$$F = \frac{\lambda_t}{\lambda_w} \beta \varepsilon_{tn} \qquad [2.100]$$

where λ_t and λ_w are thermal conductivities of tool (t) and work piece (w) materials, β is the wedge angle and ε_{tn} is the acute angle between the major and minor cutting edges.

7. The M-criterion characterizing the relative load on the cutting edges:

$$M = \frac{b_\Sigma}{b_1} \qquad [2.101]$$

where b_Σ is the total length of the cutting edges involved in cutting.

Using the seven similarity numbers we can predict such important cutting factors as components of the cutting energy, cutting temperature, residual stresses, wear volume and optimum cutting speed. For example, in order to establish the optimum cutting speed v_o, the following general equation should be considered:

$$A = n_0 Pe^{n_1} F^{n_2} D^{n_3} E^{n_4} M^{n_5} P^{n_6} U^{n_7} \qquad [2.102]$$

where n_o-n_{1-7} are experimentally obtained constants and exponents respectively, and $P=L_w/d_w$ (L_w and d_w are the length and diameter of the work piece) and $U=B_t/H_t$ (B_t and H_t are the width and height of the tool shank) are two additional similarity numbers. In consequence, the optimum cutting speed, corresponding to optimum cutting temperature (OCT), is:

$$v_o = \frac{n_o \alpha_w}{h_1}\left(\frac{h_1 b_1 (c\rho)_w \Theta_o}{F_c}\right) \qquad [2.103]$$

Constant n_o and m_l are determined experimentally and in the tests, the cutting force F_c and Θ_o are measured simultaneously. In a log-log scale A vs. Pe, n_o=Pe and m_l=tan (slope of Pe=f(A_c) line).

2.9. References

[ARM 69] ARMAREGO E.J.A., BROWN R.H., *The Machining of Metals*. New Jersey, Prentice-Hall, 1969.

[ARM 00] ARMAREGO E.J.A., "The unifield-generalised mechanics of cutting approach – a step towards a house of predictive performance models for machining operations", *Proc. 3rd CIRP Int. Workshop on Modeling of Machining Operations*, Sydney, 20 August, 2000 p. 6-24.

[AST 99] ASTAKHOV V.P., *Metal Cutting Mechanics*, Boca Raton, CRC Press, 1999.

[BAI 75] BAILEY J.A., "Friction in metal cutting – mechanical aspects", *Wear*, vol. 31, 1975, p. 243-275.

[BOO 06] BOTHROYD G., KNIGTH W.A., *Fundamentals of Machining and Machine Tools*, Boca Raton, CRC Press, 2006.

[CHA 87] CHAKRABARTY J., *Theory of Plasticity*, New York, McGraw-Hill Book Company, 1987.

[CHA 06] CHAE J., PARK S.S., FREIHEIT T., "Investigation of micro-cutting operations", *Int. J. Mach. Tools Manuf.*, vol. 46, 2006, p. 313-332.

[CHI 85] DE CHIFFRE L., "What can we do about chip formation?", *CIRP Annals*, vol. 34/1, 1985, p. 129-132.

[CHI 00] CHILDS T.H.C., MAEKAWA K., OBIKAWA T., YAMANE Y., *Metal Cutting, Theory and Applications*, London, Arnold, 2000.

[GRZ 96] GRZESIK W., "A revised model for predicting surface roughness in turning", *Wear*, vol. 194, 1996, p. 143-148.

[GRZ 98] GRZESIK W., *Fundamentals of Machining of Metallic Materials* (in Polish), Warsaw, WNT, 1998.

[GRZ 99] GRZESIK W., "Experimental investigation of the influence of adhesion on the frictional conditions in the cutting process", *Tribol. Inter.*, vol. 32, 1999, p. 15-23.

[GRZ 03] GRZESIK W., NIESLONY P., "A computational approach to evaluate temperature and heat partition in machining with multilayer coated tools", *Int. J. Mach. Tools Manuf.*, vol. 43, 2003, p. 1311-1317.

[GRZ 04] GRZESIK W., NIESLONY P., "Physics-based modeling of interface temperatures in machining with multilayer coated tools at moderate cutting speeds", *Int. J. Mach. Tools Manuf.*, vol. 44, 2004, p. 889-901.

[GRZ 08] GRZESIK W., Advanced Machining Processes of Metallic Materials, Amsterdam, Elsevier, 2008.

[JAW 93] JAWAHIR I.S., VAN LUTTERVELT C.A., "Recent developments in chip control research and applications", *CIRP Annals*, vol. 42/2, 1993, p. 659-693.

[JAW 95] JAWAHIR I.S., "A knowledge-based approach for designing effective grooved chip breakers-2D and 3D chip flow, chip curl and chip breaking", *Int. J. Adv. Manuf. Technol.*, vol. 10, 1995, p. 225-239.

[KAT 99] KATO T., FUJII H., "Energy partition in conventional surface grinding", *ASME Trans. J. Manuf. Sci. Eng.*, vol. 121, 1999, p. 393-398.

[KLU 79] KLUFT W., KÖNIG W., VAN LUTTERVELT C.A., NAKAYAMA K., PEKELHARING A.J., "Present knowledge of chip control", *CIRP Annals*, vol. 28/2, 1979, p. 441-455.

[KOM 86] KOMANDURI R., SCHROEDER T.A., "On shear instability in machining a nickel-iron base superalloy", *Trans. ASME J. Eng. Ind.*, vol. 108, 1986, p. 93-100.

[LEE 51] LEE E.H., SCHAFFER B.W., "The theory of plasticity applied to a problem of machining", *Trans. ASME, J. Appl. Mech.*, vol. 18, 1951, p. 405-413.

[LOL 82] LOLADZE T.N., *Strength and Wear Resistance of Cutting Tools* (in Russian), Moscow, Mashinostroenie, 1982.

[MAE 95] MAEKAWA K., ITOH A., "Friction and tool wear in nano-scale machining – a molecular dynamics approach", *Wear*, vol. 188, 1995, p. 115-122.

[MAN 86] MANYINDO B.M., OXLEY P.L.B., "Modeling the catastrophic shear type of chip when machining stainless steel", *Proc. Instn. Mech. Engrs*, vol. 200, 1986, no. C5, p.349-358.

[NAK 84] NAKAYAMA K., "Chip control in metal cutting", *Bull. J. Soc. Prec. Engn.*, vol. 18, 1984, p. 97-103.

[OXL 89] OXLEY P.L.B., *Mechanics of Machining. An Analytical Approach to Assessing Machinability*, Chichester, Ellis Horwood, 1989.

[OXL 98] OXLEY P., "Development and application of a predictive machining theory", *Proc. 1st CIRP Int. Workshop on Modeling of Machining Operations*, Atlanta, 19 May, 1998, p. 35-53.

[ÖZE 06] ÖZEL T., "The influence of friction models on finite element simulations of machining", *Int. J. Mach. Tools Manuf.*, vol. 46, 2006, p. 518-530.

[REZ 81] REZNIKOV A.N., *Thermophysical Aspects of Metal Cutting Processes* (in Russian), Mashinostroenie, Moscow, 1981.

[SHA 89] SHAW M.C., *Metal Cutting Principles*, Oxford, Clarendon Press, 1989.

[SHI 93] SHIMADA S., TANAKA H., OHMORI G., UCHIKOSHI J., "Feasibility study on ultimate accuracy in microcutting using molecular dynamics simulation", *CIRP Annals vol.* 42/1, 1993, p. 91-94.

[STE 97] STEPHENSON D.A., AGAPIOU J.S., *Metal Cutting Theory and Practice*, New York, Marcel Dekker, 1997.

[TOP 03] TOPOROV A., KO S.-L., "Prediction of tool-chip contact length using a new slip-line solution for orthogonal cutting", *Int. J. Mach. Tools Manuf.*, vol. 43, 2003, p. 1209-1215.

[TRE 00] TRENT E.M., WRIGHT P.K., *Metal Cutting*, Boston, Butterworth-Heinemann, 2000.

[TUR 82] TURLEY D.M., DOYLE E.D., "Microstructural behaviour – its influence on machining", in *On the Art of Cutting Metals – 75 Years Later*, PED-vol. 7, ASME, 1982, p. 99-118.

[USU 78] USUI E., HIROTA A., "Lee-Shaffer's type solution for oblique cutting", *Bull. Japan Soc. Prec. Eng.*, vol. 12, 1978, no. 2, p. 72-110.

[VIN 85] VINOGRADOV A.A., Physical Foundations of the Process of Drilling Difficult-to-Machine Materials using Carbide Drills (in Russian), Kiev, Naukova Dumka, 1985.

[VYA 99] VYAS A., SHAW M.C., "Mechanics of saw-tooth chip formation in metal cutting", *Trans. ASME J. Eng. Ind.*, vol. 121, 1999, p. 163-172.

[WEB 86] WEBER H., LOLADZE T.N., *Grundlagen des Spanens*, Berlin, VEB Verlag Technik, 1986.

[ZOR 66] ZOREV N.N., *Metal Cutting Mechanics*, Oxford, Pergamon Press, 1966.

Chapter 3

Finite Element Modeling of Machining Processes

3.1. Introduction

Although many improvements have been made during the last several years on the knowledge of cutting using analytical, mechanistic and experimental modeling approaches, there are still many issues that remain not well understood, especially with complex tool-work piece engagement conditions such as machining-induced residual stresses, stress load and distribution on the cutting tool, tool wear mechanisms and fracture, sophisticated mechanism of chip formation, etc. [VAN 98].

Finite element method (FEM)-based numerical modeling can provide a more comprehensive and, in some cases, complementary approach to experimental, mechanistic or analytical methods to virtually study machining processes in a computer environment [Klo05, Öze08, Att08]. It offers the capability to predict what could happen during the material removal process, and it could thus be possible to design and modify the process input parameters beforehand in order to reduce or eliminate problems that may arise during actual machining operations.

Chapter written by Pedro J. ARRAZOLA and Tugrul ÖZEL.

In addition, it can help in adaptive manufacturing systems, providing the possibility of correcting the process quicker when a problem appears, i.e. if a process model has been set-up, the influence of input parameters on output results should be well known, and thus corrections in process input parameters can be done faster and not by a trial and error approach.

In material removal processes, all the input parameters can have a remarkable influence on attained results: machine-tool, coolant, tool-holder, cutting tool or tool insert, tool clamping, etc. (Figure 3.1). However, considering the whole machining system, it seems quite complex to solve one given problem. Therefore, in many cases the area or elements that are going to be analyzed in depth are isolated from the rest of the system. For instance, when analyzing the surface integrity, we isolate the area close to the cutting edge (nose of the tool insert and a small area of the work piece) from the rest of the system as it is considered that their influence would not be so important.

Depending on the problem to be studied, FEM of machining processes can be focused on different scales (Figure 3.2):

– macroscale: models that will take into account the machine-tool interaction and the entire work piece and tool engagement, including their workholding and toolholding systems [Mas04, Rat05];

– mesoscale: models which consider the area where the chip is formed and a section of the work piece machined to be able to predict cutting forces, residual stresses, tool temperatures and stresses [Pie05, Klo05, Li06, Aur06, Uhl07, Öze08, Att08];

– microscale: models that will study the influence of the microstructure, grain boundaries and crystalline plasticity during chip formation [SIM 07].

While in the meso and microscales, the need for a fully coupled thermo-mechanical analysis is quite clear (even combined with a chemical analysis), at the macroscale an analysis of thermo-mechanical (but not necessarily fully coupled) could be sufficient.

This chapter will focus on the modeling approaches at mesoscale, i.e. what happens at the cutting edge.

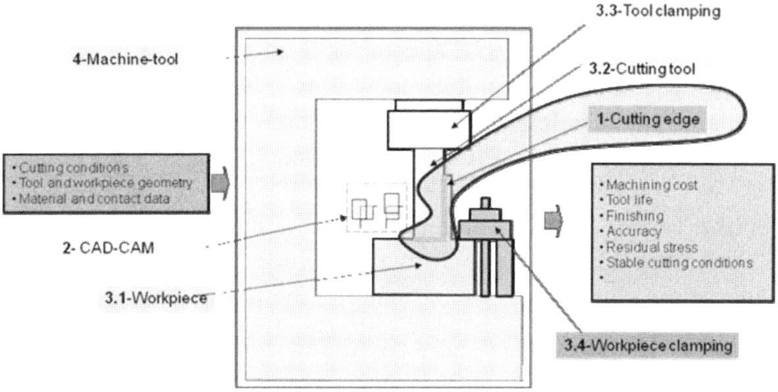

Figure 3.1. *Elements to be considered in the machining process*

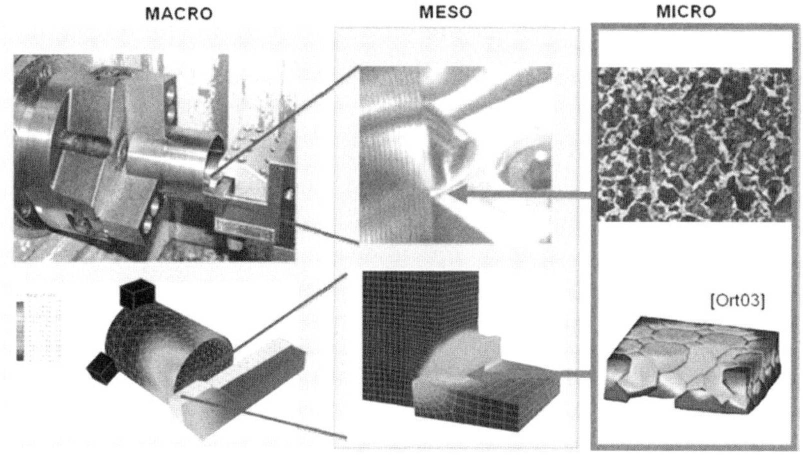

Figure 3.2. *Scales in modeling machining operations*

3.2. Finite element modeling at the meso scale (cutting edge)

Since the 1970s, there have been several applications of FEM of chip formation processes [Oku71, Tay74, Tay76]. At first, custom made codes were generated to simulate cutting processes. Later, however, most of the development was done in general-purpose software like ABAQUS™, NIKE2 (forging), ALGOR™ (metal forming), FORGE2D™, DEFORM2D™, MARC™, LS DYNA™, FLUENT™ [NG 00]. Two specific pieces of commercial software have appeared in the market during the last decade: AdvantEdge™ and Deform™. It is worth mentioning that new numerical methods ("meshless technique" [CHI 01], X-FEM, P-FEM Molecular Dynamics, etc.) have appeared during the last few years in order to solve some problems in FEM like, for instance, mesh distortion.

Unlike earlier chip formation cutting models [Mer44, Lee51, Arm69, Dud97, Alt00], FEM of chip formation processes [Leo99, Pan96, Mad00] provides some advantages due basically to the following aspects:

– access to fields of values of thermo-mechanical variables such as plastic strain, temperatures, strain rate, stresses, residual stresses, etc. that are hardly measured through experimental tests, with an interesting ratio of computational cost per information precision. It thus helps to better understand what happens during the chip formation process;

– it allows us to consider the non-linear effects of the friction at the tool-chip interface, as well as the influence of the plastic strain, strain rate and temperature, in the material behavior law;

– it is possible to realize virtual machining of tests that are difficult to justify experimentally: new tool geometries and materials or coatings, study cases that are difficult to realize (zero friction coefficient, materials with ideal behavior, etc.). To sum up, it could give ideas direction for the development of new cutting tools, cutting parameter selection, new work piece materials development, etc.;

– compared with empirical modeling it is more flexible, allowing an easy change of input parameters.

However, at this stage, it can be said that basically three kinds of problems can be found that relate to numerical cutting modeling, prior to becoming a reliable tool for industry (see Figure 3.3):

– finite element model definition;

– finite element model validation;

– identification of finite element model input parameters.

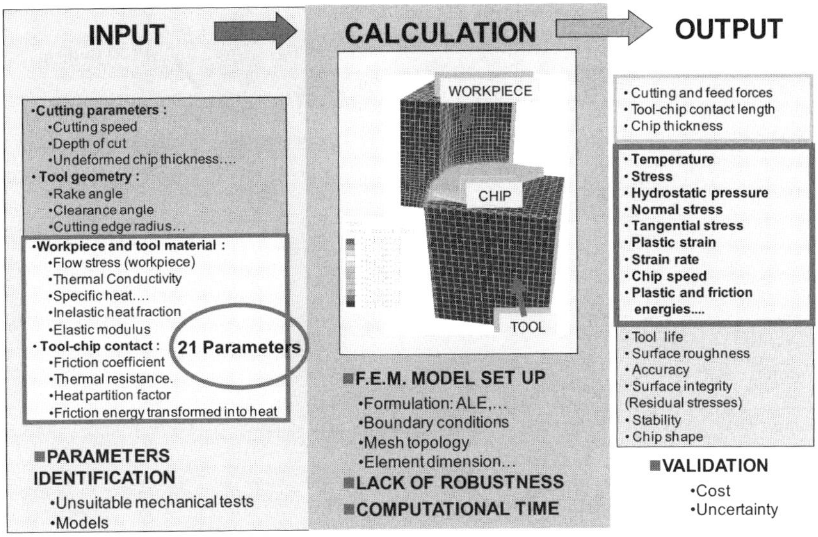

Figure 3.3. *FEM of chip formation process*

The thermo-mechanical model of the chip formation process only takes into account one area of the part and the tool, i.e. the area where the chip is formed, as shown in Figure 3.4 [ARR 03b]. In most of the research work done up until now, instead of a 3D model, a 2D model is considered, i.e. say orthogonal cutting, where it is supposed to be in a plane stress case. Even if this approach is a restriction from a machining point of view (often a 3D case), it is considered accurate enough in many cases to give an overview of what is happening. Furthermore, it allows a reduction of computational time.

However, it should be clear that in 2D modeling of many machining operations: turning, drilling, milling, etc. (with exception of broaching, sawing, etc.), the final surface obtained does not correspond to the final obtained in 3D. Thus, predicting some issues like residual stresses cannot have any realistic meaning.

One of the 2D machining cases which involves orthogonal cutting is the end-tube turning operation, which is shown in the upper side of Figure 3.4.

It is worthwhile mentioning that during recent years, several research works have been developed in 3D [Pie05, Klo05, Li06, Aur06, Uhl07, Sim07, Öze08, Att08] showing the remarkable future possibilities of FEM.

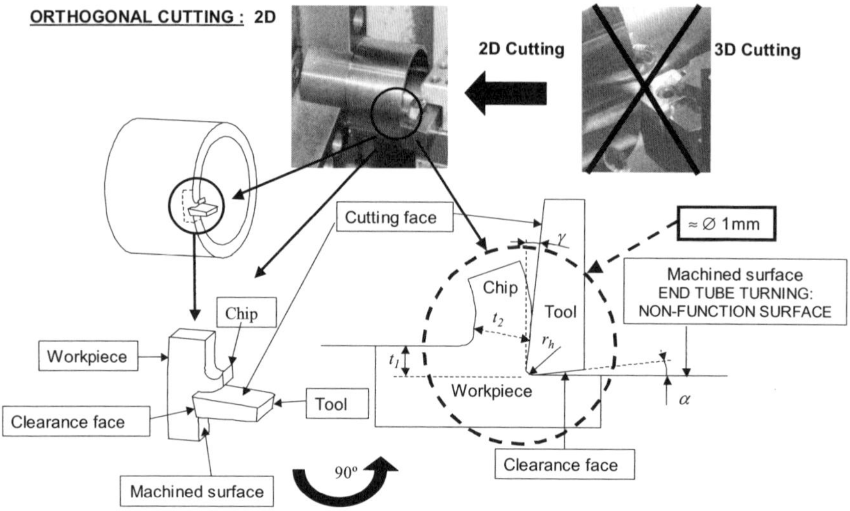

Figure 3.4. *Studied area in numerical cutting modeling [ARR 03b]*

3.2.1. *Finite element models: formulations*

Considering the research work done up until now, it is possible to establish a classification according to the formulation types: Lagrangian, Eulerian or Arbitrary Lagrangian Eulerian (ALE).

3.2.1.1. *Lagrangian formulation*

The Lagrangian description is used mainly in the solid mechanics. The domain is discretized with a mesh that follows the material at every moment; consequently, nodes move simultaneously with the latter (see Figure 3.5). Thus, the integration points are always fixed to the material. The element borders and the nodes remain the same during the evolution of the calculation, which greatly simplifies the application of the boundary conditions. At this stage, the main problem comes from the mesh control. During the simulation it should stand against the strong distortions that appear during the chip formation. Two solutions are generally used:

– the uncut chip portion of the work piece is "glued" over the work piece final surface.

The future chip glued initially to the final surface, separates from the work piece body when a geometric (distance of the cutting edge to the fracture) or mechanic (it reaches a certain value of plastic strain) criterion is reached [Str85, Ng02, Ued93, Shi03, Mar95, Söh01]. In order for the chip to be formed, once some of the mentioned criteria are reached, two techniques are used basically: the elimination or erosion of elements, or separation of the chip by the nodes that are glued to the work piece. These methods present several disadvantages:

(i) the location of the final surface of the piece is already predicted from the beginning;

(ii) some aspects where the cutting edge play an important role, are not analyzed correctly (plastic strain and residual stresses on the machined surface);

(iii) need to identify additional parameters regarding the criterion of separation of chip formation.

Figure 3.5a shows an example of a temperature field obtained during the milling of AISI 1045 steel at the cutting speed of 500 m·min^{-1}, with coated tool inserts grade P25. The author [SÖH 01] employs the method of element erosion to generate the chip (in those elements that reach a given value of plastic strain);

– remeshing: when the distortions of the work piece mesh are extremely high, a new mesh is generated [Mar95, Mad99, Fou99, Mad00, Cer96, Cer99, Cer00, Cer01, Mar02]. After a mapping of the data from the distorted mesh to the new one, the calculation continues until the criterion that determines the regeneration of the mesh is reached again. The criterion that promotes the mesh renovation can be, for example, the plastic strain. Figure 3.5b shows an example of the temperature field obtained with the commercial program AdvantEdge v4.2, when machining in 3D cutting conditions a steel AISI 4140 at the cutting speed of 300 m·min^{-1} with an undeformed chip thickness of 0.2 mm. Temperatures of close to 900°C are reached (1,200 K).

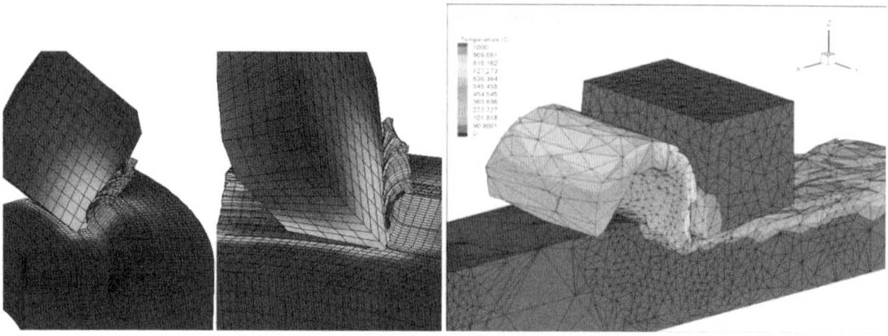

(a) (b)

Figure 3.5. *Examples of FEM with Lagrangian formulations: (a) separation line to form the chip during milling. Temperature field obtained during milling of AISI 1045 steel with coated tool insert (grade P25) at the cutting speed of 500 m·min^{-1} [SÖH 01]; (b) remeshing to form the chip. Temperature field obtained during orthogonal cutting of AISI 4140 steel, at the cutting speed of 300 m·min^{-1} [MAR 02]*

Serrated chip can be predicted provided that special criteria, which are difficult to fit in terms of physical parameters [Rec64, Kom82, Bar02, Guo03], are provided. Finally, if the steady state is searched, it is necessary to consider a very long calculation time, especially if the element size is small.

3.2.1.2. *Eulerian formulation*

The Eulerian description considers a fixed region in the geometric space occupied by continuum material examined on the time. This approach is used mainly in the fluid mechanics. The space domain is discretized and the movement must be determined with a mesh that stays fixed at every moment, i.e. the nodes do not move during the analysis and they are not fixed to the material (see Figure 3.6).

Figure 3.6a shows an example of the stress field obtained with a Eulerian formulation in a program developed by Leopold [Leo98, Leo99]. The borders of the elements and the nodes do not match with the contour at every instant and therefore the application of the boundary conditions must be realized on points that are not the nodes. This creates difficulty in the calculations [Car88, Str90, Str97, Leo99, Mae01]. This approach is very far from being executed, but great difficulties appear when dealing with the free surface treatment. Indeed, as in the beginning the simulation of the chip shape is not known, algorithms of adaptation of the mesh will be needed.

As in the case of the Lagrangian formulation, to predict the serrated chip formation, a crack propagation criteria is needed. On the other hand, this approach is used solely in the steady state analysis.

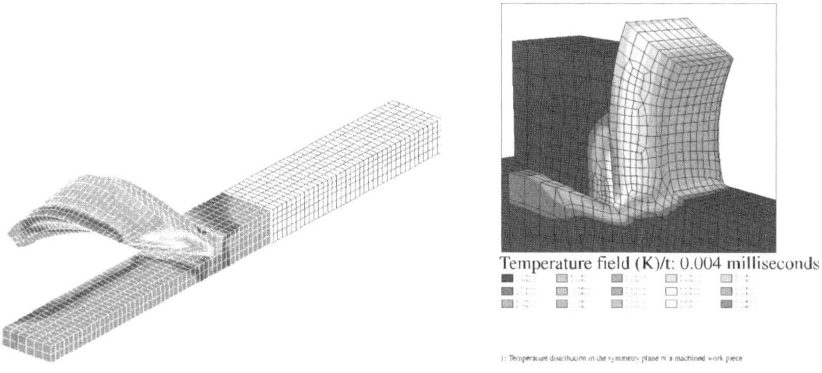

(a) (b)

Figure 3.6. *(a) Eulerian formulation. Von Mises stress field during oblique turning [LEO 99] (b) Arbitrary Lagrangian Eulerian Formulation (ALE). Temperature field obtained during machining of an AISI 4140 at 4 m·s^{-1} using the software Radioss [Pant96]*

3.2.1.3. *ALE (Arbitrary Lagrangian Eulerian) formulation*

Another method of description of the movement is the ALE Formulation or mixed description [Pant96, Mad99, Mov00, Arr07]. In this formulation, the nodes can move, but always remain inside the boundary region defined in the ALE domain (Figure 3.6b). Inside this domain, the nodes move until the distortion of the mesh diminishes. This way, the major disadvantages of the previous formulations are eliminated, conserving their advantages: easy application of the boundary conditions, easy treatment of the interfaces and lower distortion of the mesh.

These advantages make the ALE formulation suitable for the analysis of forming and crash where remarkable distortions take place.

Figure 3.6b shows the temperature field obtained when machining in orthogonal conditions of AISI 4140 steel at 4m·s^{-1} (240m·min^{-1}). It can be observed that the temperature reached is close to 1180 K in the chip and 950 K at the tool rake surface.

Computing times are smaller than in the previous formulations. It allows transitory and steady state analysis without the need to readapt the chip geometry when it is deformed. Using a small element dimension, the segmented chip can be predicted [ARR 07a] (see Figure 3.7a). Several references of this formulation can be found in the bibliography and even 3D applications [ARR 07b]; see Figure 3.7b.

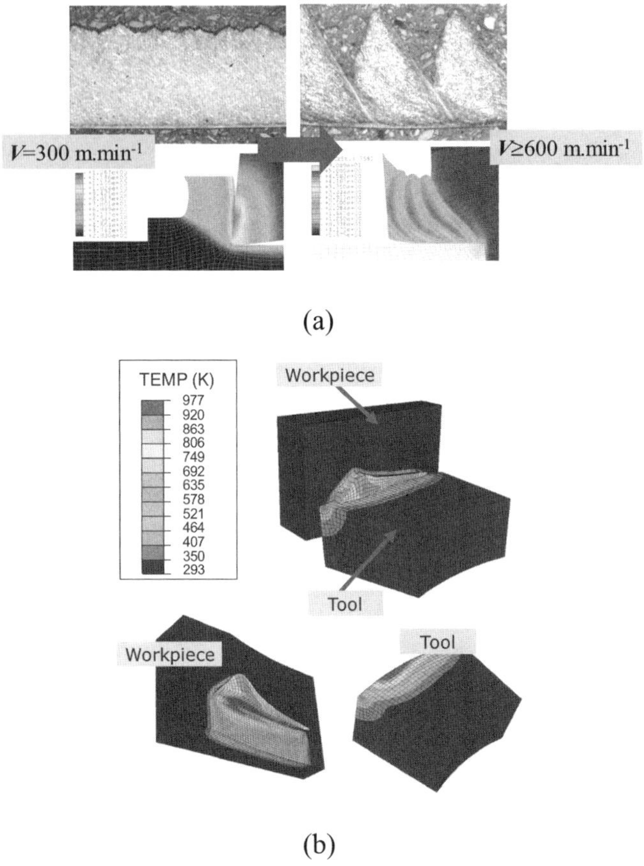

Figure 3.7. *(a) Serrated chip obtained with Abaqus/Explicit when machining an AISI 4140 steel at the cutting speed of 600 m·min^{-1}for FEM results and at a lower speed for experimental results [ARR 07a]; (b) 3D finite element model setup in Abawus Expliict with an ALE formation. Temperature field is observed when machining Inconel 718*

3.2.2. *Commercial software*

The two major pieces of commercial software existing in the market are AdvantEdge™ and Deform™. Both pieces of software provide an interface to the end user, in order to ease the introduction of process parameters and, in a way, make transparent those subjects dealing with the mathematical theory of finite element method.

Deform™ is software specialized for modeling machining operations in 2D and 3D based on an implicit integration method with fully coupled thermomechanical analysis. Since the formulation is Lagrangian, adaptive remeshing technique is used to reduce the mesh distortions when the chip is formed.

Several material constitutive laws can be used to model the material behavior: Oxley's equation and Johnson-Cook equation among others. A material database can be found in this software, both for tool and work piece materials. For modeling the contact at the tool/chip interface, a constant shear factor friction law or Columb friction law can be incorporated.

Four-node tetrahedral finite elements are used for the tool and work piece discretization.

DEFORM™ 3D machining can simulate different machining operations: turning, boring and drilling.

Work piece and tool geometries should be configured by the user, both in terms of the external dimensions as those of the mesh of the two parts. As for the tool, it should be noticed that in this software the cutting edge radius of the tool is not taken into account. However, coating layers can be implemented in the tool. It is also remarkable that the availability of loading some existing geometries of tools and toolholders from a database are incorporated in the program.

AdvantEdge™ is an explicit, dynamic, fully thermo-mechanically finite element software specialized for modeling machining operations in 2D and 3D.

Lagrangian techniques are employed for modeling the metal cutting process. Tracking discrete material points and using a predetermined line of separation to generate a fictitious crack, is possible to obtain the chip at the work piece. As the formulation is Lagrangian, an adaptive remeshing technique is used to reduce the mesh distortions.

In AdvantEdge, material behavior of the work piece is modeled by the Marusich and Ortiz constitutive model. This model takes into account deformation, strain rate and temperature to establish the flow stress. A material database can be found in this software, both for tool and work piece materials. For modeling the contact at the tool/chip interface a stick-slip Coulomb friction law is incorporated.

In 2D models the software uses six-node quadratic triangular elements and in 3D models four-node tetrahedral finite elements are used.

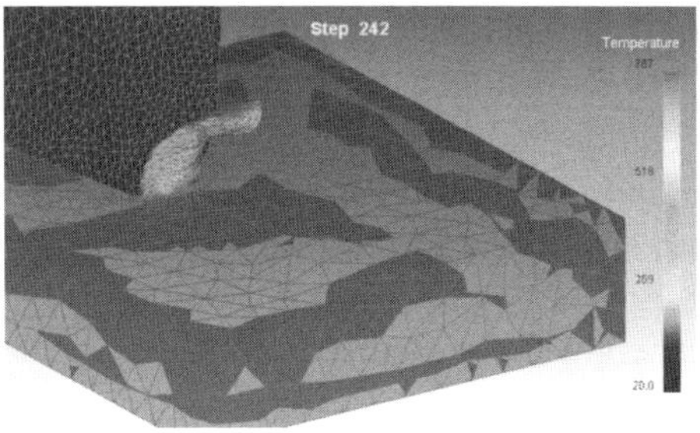

(a)

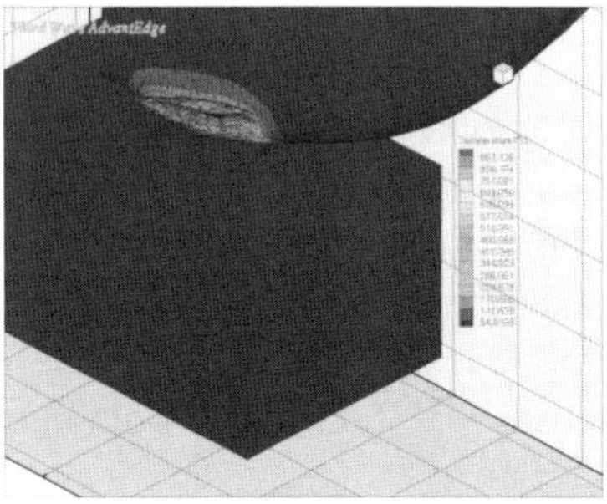

(b)

Figure 3.8. *(a) Deform 3D™ temperature field obtained in a nose turning operation;(b) Thirdwave System™ temperature field obtained in a nose turning operation*

AdvantEdge can simulate different machining operations:

– in 2D: turning, up and down milling, sawing and broaching;

– in 3D: turning, milling, tapping, grooving, drilling and boring. Furthermore, in 3D operations, it is possible to define many typologies and tool configurations (solid tool, indexable tool, etc.) as well as other operation parameters.

Tool and work piece geometries can be configured by the user both in terms of the external dimensions and the mesh of the two parts. It should be noted that, in this software, the cutting edge radius of the tool, coatings and wear phenomena can be modeled, and it brings the possibility of creating dynamic tools with vibrations, vertical velocities or damping. The availability of loading geometries of commercial cutting tools from a database incorporated in the program should also be noted. There is also a possibility of modeling composite materials.

Specific process characteristics can be defined as well, like the use of coolant and its properties, the friction coefficient at the contact or the option to predict burrs in the work piece.

3.3. Sensitivity study to define input parameters

A rigorous sensitivity study can give a good idea of the input parameters that have the highest influence on FEM results and, in a way, those that need priority in their identification. Unfortunately, commercial software are not suitable to modify all input parameters easily. Thus, general purpose proven FEA software such as Abaqus, which allow the user to model the chip formation process and easily modify entry data for the same model, are preferred in sensitivity studies.

3.4. Finite element model definition

The FEM used in sensitivity studies will be discussed here. In Figure 3.9 the scheme of the 2D numerical model set up in Abaqus/Explicit 6.1, that uses the adaptive meshing option (Arbitrary Lagrangian Eulerian formulation) is shown.

The CPE4RT element type was employed in both the work piece and the tool.

The well-known Johnson-Cook formulation is often used to define the thermo-visco-plastic behavior of the work piece material while the tool is considered to be a rigid body.

The Columb friction model governs the mechanical contact at the tool-chip interface.

The computational time (INTEL Pentium IV processor running at 2 GHz with 1 Gb of RAM) for the reference model (3 milliseconds) is close to 36 hours. This calculation time will depend very much on the element size. If smaller calculation times are expected, one solution could be the use of mass scaling techniques (for explicit integration frame).

Entry parameters, related to material and contact, and those related to process and numerical parameters are shown in Table 3.3. Column 2 of Table 3.3 shows a reference value given for all of them. They will be employed in the numerical model when the concerned parameter is not analyzed in the sensitivity study detailed in section 3.3.

A more detailed description of the model can be found in [Arr03a, Arr03b, Arr07, Arr08].

2D analysis is a restrictive approach from an industrial point of view but is considered accurate enough to make the sensitivity analysis of the numerical results to input parameters. It also allows us to reduce the computational time.

Figure 3.9 shows the temperature field over the work piece, chip and tool and after the steady state is reached. This is after 3 milliseconds of machining time, which means 15 mm of cut length at the cutting speed of 300 $m.min^{-1}$. In/out parameter values are those shown in the column called "reference value" in Table 3.3. The maximum tool temperature reaches about 1,267 K. It is located far from the cutting edge [TRE 91] and approximately at the distance of the undeformed chip thickness (t1).

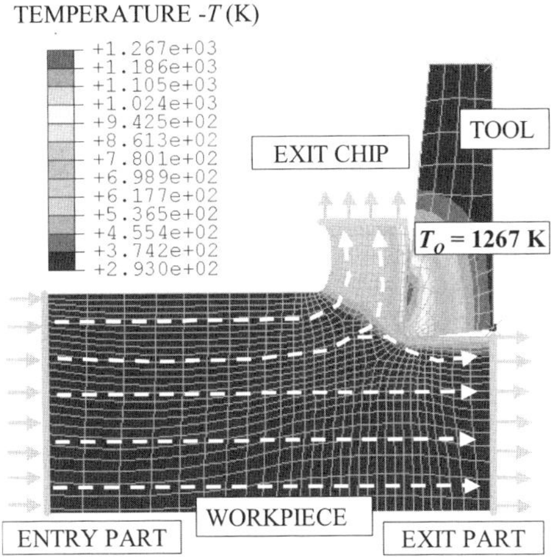

Figure 3.9. *2D numerical model set up (Abaqus/Explicit 6.1). Temperature field for reference value parameters [Arr03a, Arr03b]*

3.5. Sensitivity analysis of input parameters

3.5.1. *Test plans*

The number of parameters considered, regarding the material and contact behavior, is 14. This means that a factorial design at two levels would need 214 numerical tests, i.e. 16,384 tests. Bearing in mind calculation time, a decision to divide all entry parameters, related to material and contact behavior, into four factorial designs is taken (see Figure 3.10).

The major shortcoming of this approach is that interactions among parameters from different factorial designs will not be taken into account.

Figure 3.10 shows the parameter values for each level in the factorial design. They have been chosen according to values found in [MH178, MH380, Bro96] and take industrial applications into account. It is considered that the chosen values represent those corresponding to metals.

Obviously, the experimental validation of such a plan is impossible nowadays because this would require two work piece materials with, for example, different conductivities but identical mechanical properties.

Fac. designs	Parameters
1-Material flow stress	Yield strength: A
	Hardening modulus: B
	Strain rate sensitivity coefficient: C
	Strain hardening coefficient: n
	Thermal softening coefficient: m
2-Material thermal	Work piece material conductivity: k_p
	Work piece material specific heat: c_p
	Tool material conductivity: k_o
	Tool material specific heat: c_o
3-Tool-chip contact	Thermal conductance: K_i
	Heat partition coefficient (heat to chip): Γ
	Friction coefficient: μ
4-Percentage of energy transform. into heat	Inelastic heat fraction: β
	Friction energy transformed into heat: η

Figure 3.10. *Factorial designs in the sensitivity study*

Figure 3.15 shows numerical effects of all entry parameters over variables like the maximum tool temperature (T_o), the maximum work piece Von Mises stress (σ_{eq}), the maximum plastic strain (ε^{pl}), the tool-chip contact length (h), the chip thickness (t_2), the cutting force (F_c) and the feed force (F_f).

3.5.2. *Analysis of the results*

3.5.2.1. *Material parameters*

As can be observed in Table 3.3, although the ratio between the maximum and minimum values is not always the same, it can be seen that the Johnson-Cook flow stress coefficients *A*, *B*, *C* and *m* have a large influence over several exit parameters: the maximum tool temperature (T_o), material Von Mises stress (σ_{eq}), cutting force (F_c), feed force (F_f), chip thickness (t_2), etc.

As those coefficients increase, the material flow stress increases as well, and so will thermo-mechanical variables such as normal stress over the rake surface, the cutting forces, chip thickness, the temperature at the tool rake face and so on. It seems a logical tendency: as the flow stress increases, more force and energy is needed to form the chip. It is observed that a variation of thermal softening exponent m from 0.1 to 2 increases in the cutting force (F_c) by 120%, in the feed force (F_f) by 100%, in the maximum temperature over the rake surface (T_o) by 70% , in the Von Mises stress (σ_{eq}) by 89%, etc., until the minimum influence is over the tool-chip contact length (*h*) with 20%.

As in the case of coefficient m, it is observed that a variation of coefficient *A* from 200 to 900 increases in the maximum temperature over the rake surface (T_o) by 30%, in the cutting force (F_c) by 30%, in the feed force (F_f) by 65%, in the Von Mises stress (σ_{eq}) by 61%, etc., the minimum influence being over the tool-chip contact length (*h*) with 0%.

Analysis results of *B* showed similar conclusions to coefficient *A*. In the case of coefficient *C* it was observed that the influence was less than coefficients *A* and *B*, and of course *m*.

In the case of the hardening coefficient *n*, results should be considered with care: due to the fact that depending on the plastic strain value, for a given value of *n*, the flow strew value would increase or decrease. Regarding the interactions, it was observed that their influence was always lower than the effects of *m*, *A*, *B* and *C* coefficients.

3.5.2.2. *Material thermal parameters*

Regarding the thermal parameters, it is observed that the conductivity (k_p) and specific heat (c_p) of the work piece material have as much influence, for the chosen range, as those corresponding to the flow stress coefficients. However, in the case of the tool, these properties do not seem to have a high influence over numerical results.

3.5.2.3. *Contact parameter influence*

Among all the contact parameters, it seems that the friction coefficient has the highest influence over the numerical results. It is observed that a variation of the friction coefficient (μ) from 0.2 to 0.5 increases the maximum plastic strain over the rake surface (ε^{pl}) by 100%, the maximum temperature over the rake surface (T_o) by 40%, the chip thickness (t_2) by 23%, the cutting force (F_c) by 9%, the feed force (F_f) by 27%, etc., the minimum influence being over the Von Mises stress (σ_{eq}) with 1%.

However, all contact parameters have a large influence over the maximum tool temperature and the rake surface (T_o).

As can be observed in Figure 3.4, an important interaction appears among the three parameters analyzed over the maximum tool temperature (T_o). In the case of the heat partition coefficient (Γ) and the thermal conductance (K_i), the influence over the tool temperature (T_o) is even higher than the friction coefficient for the range of chosen values. An increase of their values in the chosen range decreases the maximum tool temperature (T_o) in -60%.

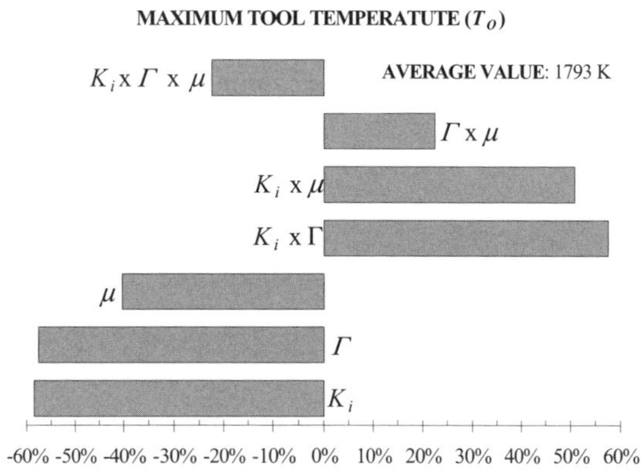

Figure 3.11. *Effects of friction coefficient (μ), heat partition coefficient (Γ) and thermal conductance (K_i) over the maximum tool temperature (T_o)*

Depending on the values of the heat partition coefficient (Γ) and the thermal conductance (K_i), the increase of friction coefficient (μ) would not necessarily increase the maximum tool temperature.

As shown in previous works [Arr03], when the friction coefficient reaches the value of μ=0.4, the chip material starts to get stuck over the rake surface and the friction energy starts to decrease while plastic energy increases. Thus, the minimum value of the maximum tool temperature (T_o) will be reached for high values of the fiction coefficient (μ), high values of the heat partition factor (Γ) and lower values of the thermal conductance coefficient (K_i).

3.5.2.4. *Inelastic and friction fraction transformed into heat parameters*

Concerning the inelastic heat fraction (β), it can be observed that, although it has an influence over nearly all the parameters analyzed, it is lower than the material and friction ones. The friction fraction transformed into heat (η) has an influence of 11% over the maximum tool temperature (T_o).

3.5.3. *Process and numerical parameter influence*

As can be observed, process and numerical parameter influence is about the same order of material and contact parameters.

Regarding numerical parameters, it seems that more work should be done in order to define a robust numerical model: mesh density and topology, basically.

Concerning process parameters, it is observed that their influence is as important as the coefficients corresponding to the material flow stress and the friction parameters.

3.5.4. *Identification tolerance*

Tool temperature being one of the most influential variables in the cutting process, the tolerance in which input parameters should be identified in order to make the maximum tool temperature vary 50 K is shown in the last column of Table 3.3. This value seems, at this stage, to be the minimum resolution temperature we can obtain after experimental tests [Dav03a,

Dav03b]. Although the model answer is not always linear to input parameter variation, these values have been extracted from the sensitivity study. It is observed that the precision of numerical results can be greatly affected by tolerance identification of input parameters.

The question that should be answered is whether there is a real capability of identifying each of these parameters with this tolerance. It is observed that, in general, tolerance for parameter identification is not given, and thus, it is not really known whether they can be reached. For example, in previous works [Arr03], tool-chip friction coefficient (μ) was identified with a ±0.02 tolerance while in the last column of Figure 3.11, a 0.03 tolerance is demanded.

3.5.5. *Inverse identification of input parameters*

Results obtained in the sensitivity study also show the non-bijectivity risk: several combinations of material flow stress coefficients can lead to the same numerical results. Taking into account the flow stress coefficients for AISI-4140 steel identified by Grolleau [GRO 96], four more virtual laws were created: Law 1, Law 2, Law 3 and Law 4. Table 3.3 shows the flow stress coefficients of these additional four virtual laws. The flow stress identified by Grolleau will be called Reference Law.

In these virtual laws, only coefficients A, n and m were modified. The selected values are far enough from the identified value, i.e., the m coefficient has been varied from m=0.62 to m=1.30, taking into account effects shown in Table 3.1 with significant variations in thermo-mechanical variables like cutting force (F_c), feed force (F_f), tool temperature (T_o), etc.

Material constitutive law	Johnson-Cook law coefficients				
	A (MPa)	B (MPa)	n ()	C ()	m ()
Reference Law	598	768	0.2092	0.0137	0.807
Law 1	400	768	0.07	0.0137	1.30
Law 2	500	768	0.15	0.0137	0.95
Law 3	700	768	0.30	0.0137	0.70
Law 4	800	768	0.35	0.0137	0.62

Table 3.1. *Material flow stress laws: reference law by Grolleau [GRO 96] and four virtual laws*

As observed, the evolution of flow stresses depending on the values of plastic strain and temperature are quite different. Numerical tests were carried out with these flow stress laws for three different cutting conditions shown in Table 3.1 while keeping the rest of the input parameters constant: reference cutting conditions, cutting conditions 1 and 2.

Figure 3.12 shows the temperature field on the work piece, chip and tool, obtained when using the material flow stress parameters corresponding to the Reference Law and Laws 1 and 2, for reference cutting conditions.

As can be observed, the results are quite similar regarding the maximum tool temperature and the temperature distribution. The differences between the highest and lowest value are lower than 7 K when cutting conditions are kept constant.

In all cases the temperature distribution is quite similar, the highest temperature value being located far from the cutting edge at the distance of approximately one and a half times the undeformed chip thickness (t_1).

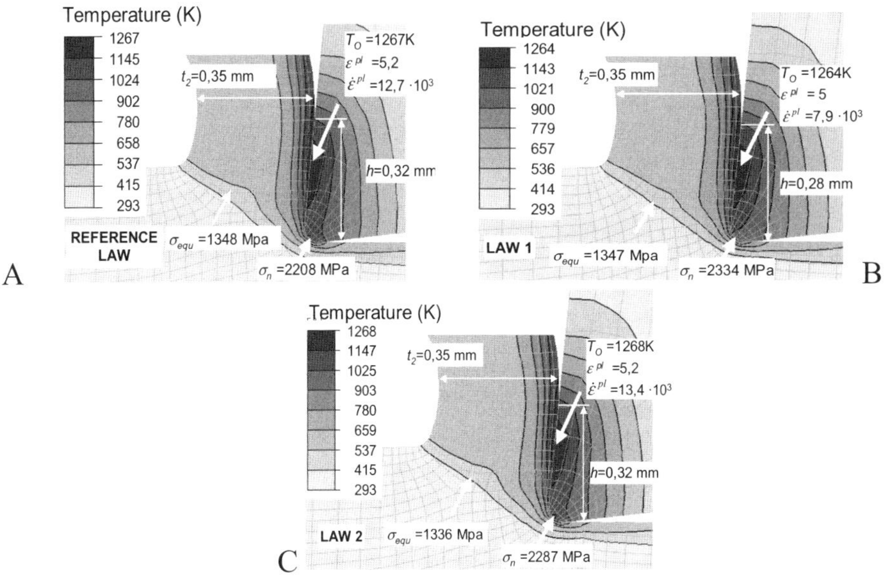

Figure 3.12. *Temperature field with the Reference Law (A), Law 1(B) and Law 2(C), when reference cutting conditions are employed*

As is observed in Table 3.2, numerical results obtained for other thermo-mechanical variables like the stress (σ_{eq}), cutting force (F_c), feed force (F_f) and so on, were also quite similar for the four material laws employed, when cutting conditions were kept constant. Differences of less than 2% were found in nearly all the thermo-mechanical variables, except in the case of the normal stress σ_n which varied 6%. These differences, as in the case of the temperature, cannot be discerned in experimental tests at this stage. For instance, and as it has been said before, the maximum tool temperature resolution data found in the bibliography is ±25K [DAV 03b].

Cutting conditions	**Material constitutive law**	T_O (K)	σ_{eq} (Mpa)	ε_2^{pl} ()	*h (mm)*	t_2 *(mm)*	F_c *(N)*	F_f *(N)*
Reference cutting conditions **Cutting speed (V)=300 m·min^{-1}** **Uncut chip thickness (t_1) =0.20 mm**	**Reference Law**	1240	1348	5.2	0.32	0.35	412	135
	Law 1	1243	1347	5.0	0.28	0.35	415	137
	Law 2	1240	1336	5.2	0.32	0.35	414	135
	Law 3	1246	1343	5.1	0.32	0.36	415	132
	Law 4	1240	1353	5.1	0.32	0.35	412	133

Table 3.2. *Numerical results obtained for the Reference Law [GRO 96] AISI-4140) and the virtual laws for three different cutting conditions: cutting speed (V)=300 (m·min^{-1}); uncut chip thickness (t_1) (mm) or feed rate =0.20 (mm.rev^{-1})*

EFFECTS OF INPUT PARAMETERS UPON NUMERICAL RESULTS

☐ Less than 10% ☐ 10%-50% ☐ 50%-100% ■ More than100%

	PARAMETER		Reference Value	Range	Ratio	T_O (K)	σ_{eq} (Mpa)	ε^{pl} ()	h (mm)	t_2 (mm)	F_c (N)	F_f (N)	ΔT =50ºK
WORKPIECE MATERIAL	Plasticity Johnson-Cook Law	*A* coefficient (MPa)	598	200-900	4.5	30	61	10	0	-40	30	65	119
		B coefficient (MPa)	768	100-800	8	35	61	45	5	15	65	45	102
		n coefficient ()	0.2092	0.05–0.5	10	5	-3	25	10	45	30	-20	0.55
		C coefficient ()	0.0137	0 – 0.05	-	20	53	25	25	-20	30	45	0.01
		m coefficient ()	0.807	0.1 – 2	20	70	89	50	20	35	120	100	0.14
	Inelastic heat fraction (β)()		0.9	0.7-1	1.4	8	-4	3	-10	-10	-10	-2	0.16
	Conductivity (k) $(W{\cdot}m^{-1}{\cdot}K^{-1})$	Part (AISI-4140 Steel)	42.6-33.1 (373 K-873 K)	7-180	25.7	-50	0	-32	12	12	4	-8	20
		Tool (C.C.grade P10)	25	25-100	4	-5	0	-32	3	2	2	2	25
	Specific heat $(c)(J{\cdot}Kg^{-1}{\cdot}K^{-1})$	Part (AISI-4140 Steel)	473-561 (473 K-873 K)	500-930	1.9	-25	10	-45	10	20	22	-3	100
		Tool (C.C.grade P10)	200	200-940	4.7	-5	0	-8	-2	1	1	2	400
CONTACT	Thermal conductance $(K_i)(W{\cdot}m^{-2}{\cdot}K^{-1})$		10^8	$10^3 – 10^8$	10^5	-60	0	5	0	3	0	2	$1{\cdot}10^3$
	Heat Partition coefficient (Γ)()		0.5	0.25–0.75	3	-60	0	0	0	1	0	0	0.05
	Friction coefficient (μ)()		0.23	0.2 – 0.5	2,5	-40	-1	100	23	5	9	27	0.03
	Friction energy transformed into heat (η)()		1	0.7-1	1,4	11	0	2	0	-2	0	-1	0.12
PROCESS	Cutting speed (V) $(m{\cdot}min^{-1})$		300	150-300	2	16	0	10	2	-4	0	5	/
	Uncut chip thickness (t_1) (mm) or feed rate $(mm.rev^{-1})$		0.2	0.05–0.3	6	27	0	23	133	138	130	96	/
	Cutting edge radius (r_h) (µm)		≈5	5-50	10	7	0	29	4	-2	7	56	/
	Rake angle (γ) (º)		6	-6º/+6º	/	-7	1	-23	-27	-14	-17	-74	/
NUMERICAL	Mesh density			Coarse-Fine	/	20	1	120	-15	-4	-7	36	-
	Mesh topology			Top 1-2-3	/	12	1	17	-14	-1	-4	-2	-

T_o: Maximum tool temperature over the rake surface.
ε^{pl} : Plastic strain over the chip.
F_c : Cutting force.
σ_{equ} : Von Mises stress.
h : Tool-Chip contact length.
F_a : Feed force.
t_2 : Chip thickness.
" – " before effect values means a decrease.
"/" means that the value cannot be explained in the table.

Table 3.3. *Numerical effects obtained after the sensitivity analysis with Abaqus/Explicit 6.1*

Similar conclusions were drawn when analyzing results obtained for other cutting conditions (see cutting conditions 1 and 2 in Table 3.2). The differences found for all the variables could not be discerned in experimental tests.

In summary, it is observed that, at least in some cases, different groups of input parameter values can lead to quite similar numerical results, which raises serious doubts about the possibility of parameter identification through inverse simulation as stated in [SHA 01].

Thus, when performing the initial identification of input parameters, it would be necessary to obtain their values as precisely as possible if good results are expected to be obtained through inverse identification.

3.6. Identification of input parameters

As shown in the previous section, FEM of chip formation requires the values of several parameters in order to model the different phenomena existing in the process, e.g., at least flow stress, conductivity, specific heat and inelastic heat fraction for material behavior, and friction, thermal conductance and heat partition coefficients for contact behavior. These parameters should be identified in similar conditions to those being in machining, that is high strain (1 - 4), strain-rate ($10^4 s^{-1}$ - $10^6 s^{-1}$), temperature (1000-1400K) and temperature rate (10^6-10^7 $K \cdot s^{-1}$). However, in most cases (work piece flow stress and tool-chip friction identification for example), this is a difficult task because experimental equipment is not able to reach these conditions [MAE 83] [MEY 86] [OLS 89] [JOY 94] [CHI 97] [CHI 98] [LEI 97] [MAE 99] [MES 00] [ElW01] [MEI 00] [HAM 02], etc.

A different approach is to use inverse identification. This is a combination of machining tests and numerical modeling, in order to obtain a corrected law [SHA 00] but this solution still does not result in a unique solution [ARR 03b].

3.6.1. *Material behavior law*

The parameters of the constitutive laws are obtained through the testing of material samples in experimental high speed equipment such as Taylor's impact tests or Hopkinson's compression and shear devices. As these devices cannot provide working conditions similar to those encountered in machining, the material behavior law's extrapolation to cutting conditions must be assumed to be valid. One possibility to improve the material test conditions and move closer to machining conditions is to use a specimen with a special shape [Cha01]. In this way the material is deformed at shear stresses closer to 4, strain rates of $2 \cdot 10^4$ s^{-1} and temperatures of 300ºC. These values are still far from those encountered in machining.

Johnson-Cook's constitutive model

Johnson-Cook's constitutive equation [JOH 83] is one of the most commonly used:

$$\bar{\sigma} = A + B\left(\varepsilon^{pl}\right)^n \times 1 + C \ln \frac{\dot{\varepsilon}^{pl}}{\dot{\varepsilon}_0^{pl}} \times 1 \quad \frac{T \quad T_{room}}{T_{melt} \quad T_{room}}^{m} \qquad [3.1]$$

In the above expression, ε^{pl} is the plastic strain, $\dot{\varepsilon}^{pl}$ is the plastic strain rate, $\dot{\varepsilon}_0^{pl}$ is the reference plastic strain rate (0.001s^{-1}), T the temperature, T_{amb} (293K) the room temperature, T_{fus} (1793K) the melting temperature. The coefficient A is the yield strength, B is the hardening modulus and C is the strain rate sensitivity.

Marusich and Ortiz constitutive model

Another example of constitutive law is that suggested by Marusich and Ortiz [MAR 95]. This is another law where the yield stress depends on the plastic strain, the strain rate and the temperature. However, according to the value of the speed of plastic deformation ($\dot{\varepsilon}^{pl}$) and temperature (T), the laws to be used are different. Thus it is considered that there exists a transition strain rate ($\dot{\varepsilon}_t{}^{pl}$), normally equal to 10^5 s^{-1} for steels, which imposes the use of two different sensitivity coefficients [TWS 00]:

m_1: sensitivity coefficient to the strain rate for $\dot{\varepsilon}_t^{pl} \leqslant 10^5 s^{\ 1}$.

m_2: sensitivity coefficient to the strain rate for $\dot{\varepsilon}_t^{pl} > 10^5 s^{\ 1}$.

Thus, if $\dot{\varepsilon}^{pl} \leqslant \dot{\varepsilon}_t^{pl}$:

$$\bar{\sigma} = \sigma_0 \; f(T) \quad 1 + \frac{\varepsilon^{pl}}{\varepsilon_0^{pl}}^{1/n} \quad 1 + \frac{\dot{\varepsilon}^{pl}}{\dot{\varepsilon}_0^{pl}}^{1/m_1} \qquad [3.2]$$

If $\dot{\varepsilon}^{pl} > \dot{\varepsilon}_t^{pl}$:

$$\bar{\sigma} = \sigma_0 \; f(T) \quad 1 + \frac{\varepsilon^{pl}}{\varepsilon_0^{pl}}^{1/n} \quad 1 + \frac{\dot{\varepsilon}^{pl}}{\dot{\varepsilon}_0^{pl}}^{1/m_2} \quad 1 + \frac{\dot{\varepsilon}_t^{pl}}{\dot{\varepsilon}_0^{pl}}^{1/m_1} \qquad [3.3]$$

It is also considered that the formulae *f(T),* which gives the temperature sensitivity according to Lemonds and Needleman [LEM 86], depends on the transition temperature T_{cut}. Thus, if $T \leq T_{cut}$:

$$f(T) = C_0 + C_1 \quad T + C_2 \quad T^2 + C_3 \quad T^3 + C_4 \quad T^4 + C_5 \quad T^5 \qquad [3.4]$$

where coefficients C_i need to be identified.

However, if $T > T_{cut}$:

$$f(T) = f(T_{cut}) \quad \frac{T \quad T_{cut}}{T_{fus} \quad T_{cut}} \qquad [3.5]$$

Zerilli constitutive model

Other work piece material constitutive laws such as those from Maekawa [MAE 83] and Zerilli *et al.* [ZER 87] consider that the flow stress depends on the plastic strain, the strain rate and the temperature:

$$\sigma_f = C_0 + C_1 e^{\quad C_3 T + C_4 T \ln \dot{\varepsilon}} + C_5 \varepsilon^{n_2} \qquad [3.6]$$

Maekawa constitutive law

$$\sigma = \sigma_0(\theta, \dot{\varepsilon}) \left(\int_{\theta, \dot{\varepsilon} \equiv h(\varepsilon)} e^{\frac{K}{n}\theta} \dot{\varepsilon}^{-\frac{m}{n}} d\varepsilon \right)^n \qquad [3.7]$$

where k, m and n are constants, $\sigma_0(\theta, \dot{\varepsilon})$ is a function depending on the temperature and strain rate and the integral takes into account the history effects.

As mentioned by Warnecke [WAR 02], in all the previous cases, the major drawback of all of them is that the flow stress value is obtained by multiplying the sensitivity of the material to each of the above mentioned effects: plastic strain, the strain rate and the temperature and thus, they cannot properly take into account all the interactions (see Figure 3.9b).

$$f\left(\varepsilon^{pl}, \dot{\varepsilon}^{pl}, T\right) = f_1\left(\varepsilon^{pl}, A, B, n\right) \quad f_2\left(\varepsilon, C, \dot{\varepsilon}_0^{pl}\right) \quad f_3\left(T, T_{amb}, T_{fus}, m\right) \qquad [3.8]$$

With this type of material constitutive law we will take into account the effects of inclusions or the phase changes in a material microstructure. Thus, if more precise models are desired, able to go further than the cutting forces, and for instance to make out the differences between a standard steel and another with improved machinability, these aspects will have to be taken into account.

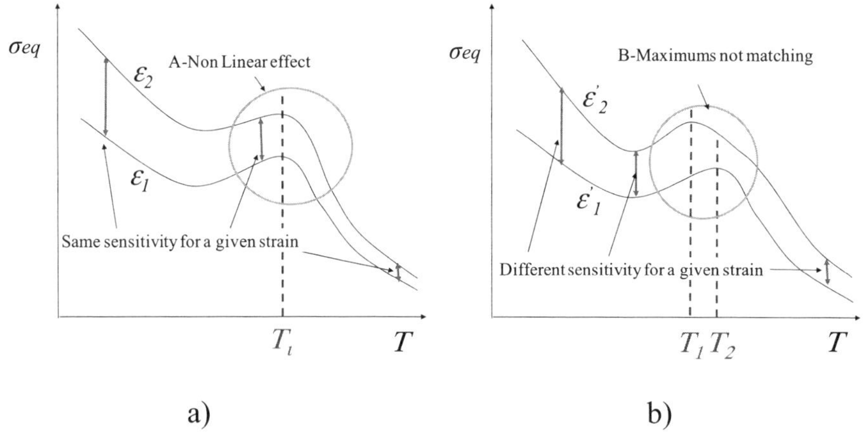

Figure 3.13. *Trend of σ depending on T according to the type of function employed: a) Johnson-Cook, Marusich, Maekawa: constant sensitivity for each variable and difficulties representing non-linear phenomena as indicated in a); b) Warnecke: possibility of representing non-linear phenomena as indicated and variable sensitivities for each variable parameter*

Warnecke [WAR 02] proposes the extraction of the law by considering that the coefficients to be identified in the law depend on the variables such as the plastic strain (ε^{pl}), the strain rate ($\dot{\varepsilon}^{pl}$) and the temperature (T).

$$\begin{aligned} f\left(\varepsilon^{pl},\dot{\varepsilon}^{pl},T\right) &= f_1\left(\varepsilon^{pl},\xi_1(\varepsilon^{pl},\dot{\varepsilon}^{pl},T),\xi_2\left(\varepsilon^{pl},\dot{\varepsilon}^{pl},T\right)...\right) \\ & f_2\left(\dot{\varepsilon}^{pl},\xi_3(\varepsilon^{pl},\dot{\varepsilon}^{pl},T)...\right) f_2\left(T,\xi_4(\varepsilon^{pl},\dot{\varepsilon}^{pl},T)...\right) \end{aligned} \quad [3.9]$$

where parameters ξ_1, ξ_2, ξ_3, ξ_4... are coefficients to be identified in material testing. This way, it is possible to consider the non-linear material behavior (see Figure 3.13a) due to the effects of inclusions or the phase changes in a material microstructure.

In some general-purpose software constitutive laws such as the Johnson-Cook law can be set up easily, while others (Maekawa, Warnecke, etc.) need to be implemented through subroutines [WAR 02].

Other drawbacks should also be considered [HAM 02] when identifying material constitutive laws:

– some interactions could not be obtained as it is not possible to perform some experimental tests: high strain and strain rates with low temperatures;

– the extreme heating rate encountered in chip formation, makes it impossible to produce a complete phase transformation, anneal softening, age hardening or even for blue brittleness effects to occur [SHI 83]. The identification of the thermal softening coefficient requires facilities that enable high temperature rate heating and temperature measurement devices. It is still possible to confuse the thermal softening and the adiabatic shear banding [CHI 98]. Some researchers have tried to use high speed heating of samples [BAS 03] [MAD 98], but the achieved heating rates (4000 $K.s^{-1}$) are far below the common machining heating rates;

– there is no standardized way of extracting coefficients provided we have the same experimental results. Thus depending on the method used, differences can be found among the coefficient identified;

– when equivalent stress states are employed, the strength differential effect [MEY 00], which could be of a certain importance in some materials, might appear and thus the same material yields different results depending on the stress state it is subjected to [GIL 00];

– the material response depends on the path followed in the strain-strain rate temperature space, as well as in the number of tests performed for the adjustment of the constituent parameters.

One of the solutions proposed is the use of a machining process to obtain flow stress data [OXL 89]. Since then, several researchers have implemented and employed different analytical models in order to obtain flow stress data [TOU 02] [SAR 04] [SHA 01] [Puj08] by means of inverse identification techniques.

3.6.2. *Friction coefficient*

Friction parameters at the tool-chip contact are hard to identify and their relevance in the cutting process is generally accepted [ÖZE 06] [FIL 07].

Several models are employed to define the behavior at the tool-chip interface.

Coulomb model

Provided that there is sliding, the tangential stress (τ) depends on the normal stress (σ_n):

$$\tau(\mathrm{x}) \leq \mu \cdot \sigma_n(\mathrm{x}) \qquad [3.10]$$

The major problem with this model is that the shear stress can be very high, even higher than the material is supposed to bear.

Sticking-sliding model [Zor63] [TRE 91]

This model, based on the Coulomb model, provides a limit value for the tangential stress which corresponds to the maximum shear stress of the work piece material τ_f:

$$\text{if } \tau < \tau_f \Rightarrow \tau(x) = \mu \cdot \sigma_n(x) \qquad [3.11]$$

$$\text{if } \tau \geq \tau_f \Rightarrow \tau(x) = \tau_f \qquad [3.12]$$

where

$$\tau_f = \frac{\overline{\sigma}}{\sqrt{3}} \qquad [3.13]$$

The major drawback of this model is that the value of τ_f, which depends on the temperature, hydrostatic pressure, etc. should be determined beforehand.

Constant shear stress model [Cer01] [SHA 00]

In this model, the normal stress is not considered. The tangential stress is a percentage of the shear stress:

$$\tau = m \cdot \tau_f \qquad [3.14]$$

where coefficient *m* can vary from 0.4 to 0.8. Again, the major drawback of this model is the identification of coefficient *m*.

Shirakashi and Usui model [SHI 74] [USU 73]

The tangential stress depends on the normal stress at the rake surface:

$$\tau = \tau_f \left[1 - e^{-K \frac{\sigma_n}{\tau_f}} \right] \qquad [3.15]$$

where τ_f is the maximum shear stress, σ_n is the normal stress and K is a constant to be determined after experimental tests for a given tool-work piece combination. In this model, the friction coefficient reaches a point where it remains constant.

Maekawa model [MAE 01]

The model takes into consideration the inclusion effects:

$$\frac{\tau}{\tau_f} = R \left[1 - e^{\left(-\mu \frac{\sigma_n}{\tau_f} \right)^n} \right]^{n'} \qquad [3.16]$$

where τ is the tangential stress, τ_f is the maximum shear stress and σ_n is the normal stress. Coefficients n, n' and R are identified from experimental tests. Coefficient R shows the average influence of inclusions over the shear stress

Model considering the temperature, pressure and sliding velocity [Sch97]

The friction coefficient depends on the sliding velocity and the normal stress over the rake surface. It considers that the friction coefficient decreases as the cutting speed (it uses sliding velocity *Vc*) increases

$$\mu = \mu(V) = \mu_0 \cdot V_c^w \qquad [3.17]$$

where μ_0 is a constant and V_c is the sliding velocity at the tool-chip interface and ϖ is a coefficient that should be identified after experimental tests.

Model considering the temperature, pressure and sliding velocity [MEI 00]

A friction coefficient depending on temperature and local normal pressure to obtain the friction coefficient values is proposed. Rech *et al.* [ZEM 07] show that the experimental friction coefficient is mainly determined by the adhesive phenomena (90%), whereas the plastic deformation is only responsible for 10% of the apparent coefficient. It obtains the following expression for the adhesive friction coefficient:

$$\mu_{adh} = 0.16 \text{ if } T > 746°C \qquad [3.18]$$

$$\mu_{adh} = 5.56.107 \text{ T -3.3008.P } 0.285\text{-}0.018 \text{ for } 545 < T < 746°C \text{ (6)} \qquad [3.19]$$

$$\mu_{adh} = 0.5 \text{ if } T < 545°C \qquad [3.20]$$

Using a device similar to Joyot, Meiller [MEI 00] we perform tests at a cutting speed close to 300 m·min^{-1}, the normal stresses (σ_n) of 730 MPa and temperatures (T) of 966ºC. In the case of the AISI 4140 V200 steel (with Calcium treatment) and a cemented carbide grade P30, it obtains the following relationship where the friction coefficient depends on the temperature, the normal stress and the sliding velocity at the tool-chip interface V_c.

$$\mu = (\sigma_n \cdot V_c) - 1\ [2 \cdot 10^{-6}\ T^3 - 0.0033\ T^2 + 1{,}953\ T - 235.65] \qquad [3.21]$$

There are only a few methods available for the identification of the friction behavior and, in all cases, experiments are not conducted in similar conditions to those encountered in the cutting process. Pin-on disk friction is done in working conditions different from those encountered in cutting; while cutting the contact of the rake surface is always with the newly generated surface, in the case of the pin-on disk, the work material (representing the work piece) that is in contact with pin is always the same. This method usually gives overestimated values.

Hedenqvist *et al.* [HED 91] proposed a new experimental set-up reaching the sliding velocity of 3 m·s^{-1}. However, the pressure was still low: 15 MPa.

Grzesik *et al.* [GRZ 02] developed an experimental set-up using a cylindrical specimen (round insert) and a flat rotating disk to measure the friction force F_f depending on the wear for the combination of a work piece and a tool material, according to the sliding distance. He investigated cutting the wear resistance of tool coatings for a high sliding speed (0.5-3 m^{-1}) and long test time (1,200 s). However, the contact pressure still remains low (≈ 10 MPa) and the pin rubs on a single track and the work piece surface is never refreshed.

The modified pin-on disk device proposed by Olsson [OLS 89] makes it possible to refresh the working material, as required during the cutting process. However, the pressures applied on the pin-on disk device are still low (25 MPa) compared with the mechanical conditions that occur during machining (close to 1,200 MPa). In the test proposed by Joyot [JOY 94] and improved by Meiller [MEI 00], the load is high enough to impose a high pressure (750 MPa). At the same time, the temperature at the interface between the frictional tool and the working material (950ºC) approaches the existing temperature during the cutting process.

Zezemi *et al.* proposed a new experimental tribometer for the characterization of the tribological phenomena in high speed dry cutting and enables us to scan a large spectrum of contact pressure up to 3 GPa, under high velocities (over 3 m/s) and high temperatures (500-800°C). The friction law is obtained combining numerical model simulations with experimental model simulations.

A different approach is to use machining tests to obtain an approximation of the Coulomb friction coefficient. An example of this idea is Albrecht's [ALB 60] or Merchant's method [MER 44]. Albrecht's work analysis [ALB 60] makes it possible to measure the tool-chip friction coefficient over the rake surface, eliminating cutting edge effects that are included when using Merchant's approach. Following Albrecht's approach, Arrazola *et al.* [Arr08] proposed a method to identify a friction law from an experimental test. They claim the need of using a variable friction coefficient (to take into account above all the cutting edge effects) and obtained after experimental machining tests to identify the friction coefficient

$$\mu_A = tg \quad arctg\frac{dF_f}{dF_c} + \gamma \qquad [3.22]$$

where

$$\frac{dF_f}{dF_c} = 0.00782 + 0.68544 \cdot 1.00215^{(772.82 \cdot h \;\; 58.132)} \;\; 0.68544 \cdot \left(1.00215^{(772.82 \cdot h \;\; 58.132)}\right) \cdot$$
$$(772.82 \cdot h \;\; 58.132) \cdot \ln 1.00215 \qquad [3.23]$$

and where h is the tool-chip contact length considered.

3.6.3. *Identification of the other parameters*

The same issues mentioned for the material constitutive law and tool-chip contact arise when considering other parameters with physical meaning and which have a remarkable influence on quantitative results. However, usually, as in the case of thermal resistance (R_i) or thermal conductance (K_i) [CER 07], these are the percentages of plastic strain transformed into heat (β) (Taylor Quinney coefficient), the percentage of friction energy transformed into heat (η), heat partition coefficient (Γ), specific heat (c), thermal conductivity (k), etc.

3.7. Finite element model validation

Obtaining the values of variables such as the cutting forces, chip geometry, speed fields, temperatures, plastic strains, strain rates, stresses within material, etc. is necessary for the validation of FEM models. The different techniques employed can be classified as techniques that provide us with information before the machining process (before-process techniques), during the machining process (in-process techniques) and after the machining process has taken place (post-process techniques).

3.7.1. *In-process techniques*

The global values of the cutting forces (without knowing the application point) in machining, with a continuous chip, are easy to determine [BOO 75] [Sha84]. Their measurement in milling remains more complicated due to vibrations and excitation of the measurement sensor. However, two normal pressure distributions can lead to the same value of the cutting force. Obtaining the normal and tangential stress measurement is always complicated. The use of photoelasticity with plexiglass tools [CHA 65] or composite tools (Split tool) have been proposed by several authors [SHI 74] [CHI 00] [MAE 01]. Difficulties arise in determining values closer to 0.1 mm from the cutting edge [MAE 01]. The use of sapphire tools makes it possible to observe and comprehend the chip formation process [KOM 89] and above all to understand the friction phenomenon.

Temperature remains one of the most important variables in machining. Several techniques have been employed up to now [BAR 73]: chip color [VEN 93] [YEO 00], thermosensitive elements, thermocouples, pyrometers, IR cameras, etc. Techniques based on the use of thermosensitive elements (varnish [Ros71] and powders [KAT 76] [Loc89]) are delicate and need several repetitions to obtain a temperature field. The method based on oxides, which change phase according to the temperature, enable us to have a temperature distribution, but with very significant space gradients.

Using tool materials which undergo microstructural transformations with the temperature, like the high speed steels, makes it possible to obtain thermal fields after the measurement of the material hardnesses [TRE 91] [WRI 78]. This technique makes it possible to obtain measurements with an accuracy of 50 K and space heat gradients of roughly 100 K per tenth of a millimeter. However, it requires the calibration of hardnesses according to the temperature and it is limited to measurements below 800ºC, i.e. it remains limited in certain cutting conditions.

Thermocouples inserted in the tool [STE 93] [LES 97] [LEC 95] present some difficulties; for instance, the knowledge of the exact localization of the thermocouple tip and that they will not be located exactly at the hottest point. Some authors take advantage of the Seebeck effect using the work piece and the tool like elements of the couple for the measurement [Bir92]. Again problems arise in defining the thermocouple's properties and in the localization of the place where the thermocouple effect occurs.

Jaspers [JAS 99] used films of very fine thickness (RDT-detectors of temperature by resistance) to measure the temperature. He even mentioned the possibility of measuring normal stresses.

The pyrometer remains an interesting technique to measure the temperature due to its relatively low cost. The measured temperature is the average value temperature on the point measured (1 mm diameter) [LEI 97]

[NG 99] [UED 01] [DAV 03a]. One of the major sticking points is the need to calibrate the emissivity, according to the temperature, wavelength, etc.

Although they are quite expensive, IR cameras remain one of the most promising solutions for temperature measurement. Emissivity calibration is also needed [MES 99b][DAV 03a][DAV 03b][ARR 08b]. Some authors use CCD-IR cameras (cameras working in the visible spectrum and using the answer close to the infra-red) [LEC 95] [MSA 98] [CHA 01] to obtain temperature fields on the chip or tool surface. They claim to obtain a spatial resolution of 0.02 mm.

A combination of high speed filming devices (up to 100,000 images a second) [WAR 77] [DEC 92] and visioplasticity [OXL 89] makes it possible to obtain plastic strain and strain rates during the chip formation process.

Quick-stop devices allow us to obtain a chip at exactly the moment where it is formed and thus, it is possible to analyze how the material flows over the cutting edge [GUI 95] [DEF 96] [SPA 71].

3.7.2. *Before and post-process techniques*

Scanning electron microscopes, optical magnifying and perfilometer devices make it possible to analyze materials and the tools before and after machining, giving ideas on the type of wear which is at the origin of the deterioration of the tools.

3.7.3. *Validation of FEM results*

An extremely precise validation of results obtained with the FEM model defined previously would have demanded the proper identification of all entry parameters, which is, as mentioned before, a difficult task as well as a time-consuming and cumbersome job. Validation was carried out comparing numerical results with experimental results and numerical results obtained from the commercial software AdvantEdge 4.3.

Data about experimental results, especially cutting forces, feed forces, chip thickness, tool-chip contact length and temperature have been obtained from experimental tests. They were carried out on machining tubes of 80 mm diameter and 3 mm wall thickness without lubrication. The work piece material was AISI-4140 steel of 262-293 HB hardness. Cemented carbide inserts of grade P10 were used as tools. In all cases, the cutting edge radius was measured giving a value of 40±5 μm.

A Kistler device (reference 9121) was used to measure the cutting forces during experimental tests. Force values are extrapolated for 1 mm depth of cut.

The commercial software AdvantEdge 4.3, compared to the model set up in Abaqus/Explicit, shows several differences: Lagrangian formulation instead of A.L.E. formulation, work piece material behavior is governed by Marusich law, specific material and contact behavior data, mesh density and topology. Thus it does not make sense to try to compare quantitative results.

In order to make a qualitative assessment rather than a quantitative one, validation has been performed comparing experimental and numerical results over a range of geometric and cutting conditions, where effects and interactions of four process parameters: cutting speed (v), uncut chip thickness (h), cutting edge roundness (r_β) and rake angle (γ_o) were analyzed in a factorial design [BOX 78]. Parameter values for each level are shown in Table 3.3 (see the range value column in the process rows). Finite element analyses are carried out with *Abaqus/Explicit*© and the commercial software for machining purposes *AdvantEdge*™ (version 4.1).

Figure 3.17 shows the temperature and stress fields when the reference values are employed in the FEM model (see the reference value column in Table 3.3).

Figure 3.18 shows the FEA and experimental effects of cutting process parameters over feed force (F_f). As can be observed, a good qualitative agreement is obtained for the three cases. For instance, it is observed that a variation of the uncut chip thickness (h) from 0.05 mm to 0.3 mm increases the feed force (F_f) by 93% in Abaqus/Explicit©, by 74% in AdvantEdge™ and by 75% in experimental tests.

Thus, the numerical model set up in Abaqus/Explicit is considered to be accurate enough to carry out a sensitivity analysis of material and contact parameters.

In the case of other variables that are compared e.g., cutting force (F_c), temperature (T_o) or plastic strain (ε^{pl}), quite good qualitative agreements were also obtained.

Therefore, notwithstanding quantitative differences between the FEA and experimental results, as outlined in Figure 3.18, the numerical model set up in *Abaqus/Explicit*© is considered to be reliable enough for a sensitivity analysis of material and contact parameters.

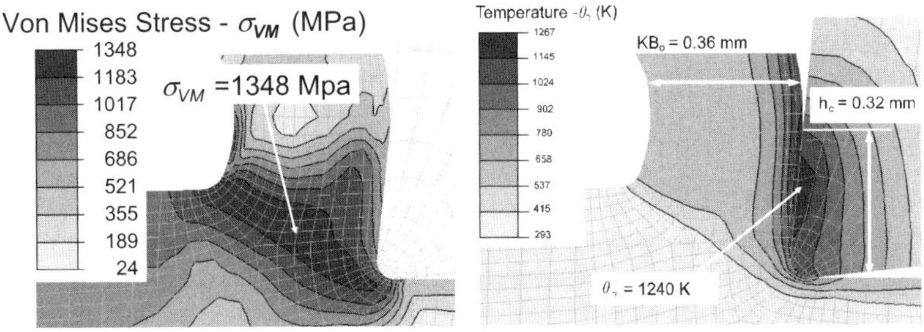

(a) (b)

Figure 3.14. *(a) Temperature and (b) Von Mises stress fields (reference values: AISI-4140 steel and P10 insert grade)*

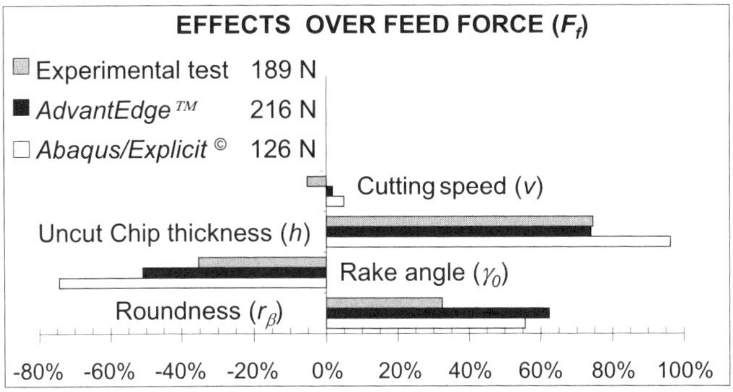

Figure 3.15. *Experimental and FEA effects over feed force*

3.7.4. *Summary*

It is observed that there are very few measurable variables: chip geometry, cutting forces, temperature, plastic strain and strain rate, etc. and the uncertainty obtained in each case seems to be non-negligible (±50K). The use of combined techniques to measure one variable (temperature: IR techniques, thermocouple, etc.) can offer a way to reduce the uncertainty.

3.8. Conclusions

The following conclusions can be drawn:

– FEM is a promising technique to study the cutting process, above all if qualitative results are desired. Values of difficult to be measure variables such as stresses, temperatures, strain and strain rates can be obtained;

– 3D modeling would be needed if industrial needs are to be accomplished;

–the sensitivity study reported here suggests the great influence of multiple parameters related to the material and contact (material flow stress, friction coefficient, thermal conductance, heat partition coefficient, work piece material conductivity and specific heat, etc.) on FEA results. The need for better identification of parameters different from those related to material flow stress and friction coefficient, e.g., thermal conductance or heat partition coefficient, is thus essential if reliable quantitative results are to be expected;

– in order to estimate the global FEM uncertainty, it is necessary to consider not only the uncertainty corresponding to the model, but also the uncertainty of entry parameter identification;

– considering the influence of numerical parameters over the numerical results, it seems that, on the one hand, a higher robustness should be desired in the model. On the other hand, quantitative results should be very carefully considered when drawing conclusions;

– even if the sensitivity study has not been validated completely, it suggests the certitude window in which some machining parameters should be adapted in order to obtain the desired qualitative values of thermo-mechanical variables;

– FEM validation results validation is a cumbersome task, considering the experimental difficulties in measuring some thermo-mechanical variables such as the temperature, plastic strain, work piece material stress, etc.

3.9. Acknowledgments

The authors would like to thank the Basque and Spanish Governments for the financial support given to the projects: CIC MARGUNE II (IE06-174), CIC MARGUNE III (IE07-204), MICROMANUFACTURING 0,0 (IE08-222) and Maquimodel (DPI2006-15502-C02-01 and PI2009-2).

3.10. References

[ALB 60] Albrecht P., "New developments in the theory of the metal-cutting process. Part1: The ploughing process in metal cutting", *Journal of Engineering for Industry*, November 1960, pp.348-358.

[ALT 00] Altintas Y., *Manufacturing Automation. Metal Cutting Mechanics, Machine Tool Vibrations and CNC Design*, Cambridge University Press, 2000.

[ARM 69] Armarengo E.J.A., Brown R.H., *The Machining of Metals*, Prentice-Hall, Inc. Englewood Cliffs, New Jersey. 1969.

[ARR 03a] Arrazola P.J., Meslin F., Marya S., "Numerical cutting sensitivity study of tool-chip contact", *Material Science Forum*, Vols. 426-432, July 2003, pp.4519-4524.

[ARR 03b] Arrazola P-J., 2003, Modélisation numérique de la coupe: étude de sensibilité des paramètres d'entrée et identification du frottement entre outil-copeau, PhD Thesis, E.C. Nantes.France.

[ARR 07a] Arrazola P. J., Villar A., Ugarte D. and Marya S., 2007, "Serrated chip prediction in finite element modeling of the chip formation process", *Machining Science and Technology*, 11:3, 367 – 390.

[ARR 07b] Arrazola P.J., Llanos I., Villar J.A., Urresti I., Ugarte D., "3D finite element modelling of chip formation process", *10th CIRP International Workshop on Modelling of Machining Operations*, Regio-Calabria, Italia. 27-28 Agosto. 2007.

[ARR 08a] Arrazola P.J., Ugarte D., Domínguez X., 2008, "A new approach for the friction identification during machining through the use of finite element modelling", *International Journal of Machine Tools & Manufacture*, 48,73–183.

[ARR 08b] Arrazola P.J., Arriola I., Davies M.A., Cooke A.L., Dutterer B.S., 2008, "The effect of machinability on thermal fields in orthogonal cutting of AISI 4140 steel", *Annals of the CIRP*, 57/1.65–68.

[ATT 08] Attanasio A., Ceretti E., Rizzuti S., Umbrello D., Micari F., "3D finite element analysis of tool wear in machining", *Annals of the CIRP*, 57/1.61-64

[AUR 06] Aurich J.C., Bil H., "3D finite element modelling of segmented chip formation", *Annals of the CIRP*, 57/147-50.

[BAR 73] Barrow G., "A review of experimental and theoretical techniques for assessing cutting temperatures", *Annals of the CIRP*, Vol.22/2/1973. pp.203-211.

[BAR 02] Barry J., Byrne G., "Chip formation, acoustic emission and surface white layers in hard machining", *Annals of the CIRP*, Vol. 51, 2002. pp.65-68.

[BAS 03] Basak D., Yoon H.W., Rhorer R., Burns T., 2003 "Mikrosecond Time-Resolved Pyrometry during Rapid Resistive Heating of Samples in a Kolsky Bar Apparatus, Temperature: Its Measurement and Control in Science and Industry", *American Institute of Physics*, Vol. 7, p.753-757.

[BIR 92] Birgand J., Mesure de température en cours d'usinage et validité de ce critère pour caractériser une nuance de matériau-outil à base de carbure de tungstène, Thesis, Ecole Centrale de Nantes, December 1992.

[BOO 75] Boothroyd G., Knight W.A., *Fundamentals of Machining and Machine Tools*, Second edition, Marcel Dekker, Inc., 1975.

[BOX 78] Box G.E.P., Hunter W.G., Hunter, J.S., *Statistics for Experiments*, John Wiley & Sons, 1978.

[BRO 96] Brookes J.A., *World Directory and Handbook of Hard Metals and Hard Materials*, sixth edition, Kenneth J A Brookes, 1996.

[CAR 88] Carroll G.J.T., Strenkowski, J.S., "Finite element models of orthogonal cutting with application to single point diamond turning", *International Journal Mechanical Sciences*, 1988, Vol.30, No.12, pp.899-920.

[CER 96] Ceretti E., Fallbohmer P., Wu W.T., Altan T., "Application of 2D FEM to chip formation in orthogonal cutting", *Journal of Material Processing Technology*, 1996, Vol.59, pp.169-180.

[CER 99] Ceretti E., "Numerical study of segmented chip formation in orthogonal cutting", *Proceedings of the 2nd CIRP International Workshop on Modelling of Machining Operations*, Nantes, France, 1999, pp.124-132

[CER 07] Ceretti E., Filice L., Umbrello D., Micari F., "ALE simulation of orthogonal cutting: a new approach to model heat transfer phenomena at the tool-chip interface", *Annals of the CIRP*, 2007, Vol. 56/1/2007. pp: 69-72.

[CHA 65] Chandrasekharan H., Kapoor D.V., "Photoelastic analysis of tool-chip interface stresses", *Journal of Engineering for Industry*, 1965, vol. 87B, pp.495-502.

[CHA 01] Changeux B., Loi de comportement pour l'usinage. Localisation de la déformation et aspect microstructuraux, Doctoral thesis, Ecole Nationale Supérieure d'Arts et Métiers, June 2001.

[CHI 97] Childs T.H.C., "Material property requirements for modeling metal machining", *Journal of Physique* IV 7 colloque C3, France, 1997, pp.xxi-xxxvi

[CHI 98] Childs T.H.C., "Material property needs in modeling metal machining", *Proceedings of the 1st CIRP International Workshop on Modeling of Machining Operations*, Atlanta, Georgia, USA, May 1998, pp.193-202.

[CHI 00] Childs T.H.C., Maekawa, K., Obikawa, T., Yamane, Y., *Metal Machining. Theory and Applications*, ARNOLD, London, Great Britain, 2000.

[CHI 01] Chinesta F., Lorong, P., Ryckelinck, D., Martinez, M.A., Cueto, E., Doblaré M., Coffignal G., Touratier M., Ivonnet J., "Thermomechanical cutting model discretisation: Eulerian or Lagrangian, mesh or meshless?", *International Journal of Forming Processes*, 2001, Volume X, pp.1-10.

[DAV 03a] Davies M.A., Yoon, H., Schmitz, T.L., Burns, T.J., Kennedy, M.D., 2003, "Calibrated thermal microscopy of the tool-chip interface in machining", *J. Machining Science and Technology*, 7/2:167-190.

[DAV 03b] Davies M.A., Cooke, A.L., Larsen, E.R., 2005, "High bandwidth thermal microscopy of machining AISI 1045 steel", *Annals of the CIRP*, 54/1:63-66.

[DEC 92] De Chiffre L., Influence of restricted contact and cutting fluid on chip curl, Technical University of Denmark, October 1992, Videotape.

[DEF 96] Deflisque L., Caractérisation d'outils céramiques. Composites base alumine en tournage de matériaux difficiles à usiner, Doctoral thesis, Ecole Centrale de Nantes, September 1996.

[DUD 97] Dudzinski D., Molinari A., "A modelling of cutting for viscoplastic materials", *Int. J. Mech. Sci.*, 1997, Vol.39, No.4, pp.369-389.

[FIL 07] Filice F., Micari S., Rizzuti D., Umbrello D., "A critical analysis on the friction modelling in orthogonal machining", *International Journal of Machine Tools and Manufacture*, 47(3-4) (2007) 709-714

[FOU 98] Fourment L., Bouchard P.O., Bay F., Chenot J.L., "Numerical simulation of chip formation and crack propagation during non-steady cutting process", *Proceedings of the 2nd CIRP International Workshop on Modelling of Machining Operations*, Nantes, France, 1998, pp.108-123.

[GIL 00] Gilat A., Cheng Ch-Sh., "Modeling torsional split Hopkinson bar tests at strain rates above 10,000 s−1", *Experimental Mechanics*, 40 (1) (2000) 54-59.

[GUI 95] Guillot D., Contribution à l'étude de l'influence inclusionnaire sur l'usinabilité des aciers bas carbone, Doctoral thesis, Ecole Centrale de Nantes, November 1995.

[GRO 96] Grolleau V., 1996, Approche de la validation expérimentale des simulations numériques de la coupe avec prise en compte des phénomènes locaux à l'arête de l'outil, PhD Thesis, E.C. Nantes.

[GRZ 02] Grzesik W., Zalisz Z., Nieslony P., 2002, "Friction and wear testing of multilayer coatings on carbide substrates for dry machining applications", *Surface and Coatings Technology*, Vol. 155, pp.37–45.

[GUO 03] Guo Y.B. "FEM simulation of discontinuous chips in machining AISI 4340 steel", *Proceedings of the 6th CIRP International Workshop on Modelling of Machining Operations*, Hamilton, ON, Canada, May 2003, pp.73-80.

[HAM 02] Hamann J.C., Meslin F., Sartkulvanich J., "Criteria for the quality assessment of constitutive equations dedicated to cutting models", *Proceedings of the 5th CIRP International Workshop on Modeling of Machining Operations*, West Lafayette, USA, 2002, pp.21-30.

[HED 91] Hedenquist P., Olsson M., 1991, "Sliding wear testing of coated cutting tool materials", *Tribol. Int.*, Vol. 23, No 3, pp.143–150.

[JOH 83] Johnson W.K., Cook R., 1983, "A constitutive model and data for metals subjected to large strains, high strain rates and high temperatures", *Proceeding of the 7th International Symposium on Ballistics*, Hague, The Netherlands, 1983, pp.541-547.

[JAS 99] Jaspers S., *Metal Cutting Mechanics and Material Behaviour*, Eindhoven University of Technology, 1999.

[JOY 94] Joyot P., 1994, Modélisation numérique et expérimentale de l'enlèvement de matière, PhD Thesis University of Bordeaux, France.

[KAT 76] Kato S., Yamaguchi K., Wanatabe Y., Hiraiwa Y., "Measurement of temperature distribution within tool using powders of contact melting point", *Journal of Engineering for Industry*, May 1976, pp.607-613.

[KLO 01] Klocke F., Raedt H.-W., Hoppe S., "3D-FEM simulation of the orthogonal high speed cutting process", *Material Science and Technology*. Vol.5, Number 3/2001.

[KLO 05] Klocke F., Kratz H., "Advanced tool edge geometry for high precision hard turning", *Annals of the CIRP*, 54/1.47-50.

[KOM 82] Komanduri R., Schroeder T., Hazra J., Von Turkovich B.F., Flom D.G., "On the catastrophic shear instability in high speed machining of a 4340 steel", *Journal of Engineering for Industry*, May 1982, Vol.104, pp.121-131.

[KOM 89] Komanduri R., Larsen-Basse J., "Tribology: The cutting edge", *Mechanical Engineering*, January 1989, pp.74-77.

[LEE 51] Lee E.H., Shaffer B.W., 1951, "The theory of plasticity applied to a problem of machining", *ASME Jour. App. Mech.*, 73:404-413.

[LEI 97] Lei S., Shin Y.C., Incropera F.P., "Material constitutive modeling under high strain rates and temperatures through orthogonal machining tests", *Manufacturing Science and Technology*, 1997, Vol.2, pp.91-98.

[LEM 86] Lemonds J., Needleman A., "Finite element analysis of shear localization in rate temperature dependent solids", *Mechanical Materials*, 1986, Vol.5, pp.339-361.

[LEO 98] Leopold J., "FEM modeling and simulation of 3-D chip formation", *Proceeding of the 1st CIRP International Workshop on Modeling of Machining Operations*, Atlanta, Georgia, USA, May 1998, pp.235-245.

[LEO 99] Leopold J., Schmidt G., 1999, "Challenge and problems with Hybrid Systems for the modeling of machining operations", *Proc. II CIRP Inter. WMMO*, 298-311.

[LES 97] Leshoc C.E., Shin Y.C., "Investigation on cutting temperature in turning by a tool-work thermocouple technique", *Journal of Manufacturing Science and Engineering*, November 1997, Vol.119,pp.502-508.

[LEC 95] Le Calvez C., Etude des aspects thermiques et métallurgiques de la coupe orthogonale d'un acier au carbone, Doctoral thesis, ENSAM de Paris, December 1995.

[LI 06] Li R., Shih A.J., "Finite element modeling of 3D turning of titanium", *International Journal of Advanced Manufacturing Technology*, 2006, Vol 29, n 3-4, pp. 253-261.

[MAD 99] Madhavan V., Chandrasekar S., Farriss T.N., "Machining as a wedge indentation", *Proceedings of the 2nd CIRP International Workshop on Modeling of Machining Operations*, Nantes, France, 1999, pp.298-311.

[MAD 00] Madhavan V., Gandikota V.A., and Agarwal R., "Non-linear finite element analysis of machining and sheet metal forming", *AIAA Journal*, 2000, Vol.38, No.11, pp.2176-2186.

[MAD 98] Macdougall D., "A radiant heating method for performing high-temperature high-strain-rate tests", *Meas. Sci. Technol.*, 1998, Vol. 9, pp. 1657-1662.

[MAE 83] Maekawa K., Shirakashi T., Usui E., 1983, "Flow stress of low carbon steel at high temperature and strain rate (Part 2)", *Bull. Japan. Soc. Of Prec. Engr.*, 17/3:167-172.

[MAE 99] Maekawa K. "Application of 3D machining modelling to cutting tool design", *Proc. of the 2nd CIRP Int. WMMO*, Nantes, France, 1999, pp.206-233.

[MAE 01] Maekawa K., Kubo A., Childs T.H.C., "A friction model for free-machining steels and its applicability to machinability analysis", *Key Engineering Materials*, Vol.196, 2001, pp.79-90.

[MAR 95] Marusich T.D., Ortiz M., "Modeling and simulation of high-speed machining", *International Journal for Numerical Methods in Engineering*, 1995, Vol.38, pp.3675-3694.

[MAR 02] Marusich T.D., Brand C.J., Thiele J.D., "A methodology for simulation of chip breakage in turning processes using an orthogonal finite element model", *Proceedings of the 5th CIRP International Workshop on Modeling of Machining Operations*, West Lafayette, IN, USA, May 2002, pp.139-148.

[MAS 04] Masset L., Debongnie J-F., "Machining processes simulation: specific finite element aspects", *Journal of Computational and Applied Mathematics*, v. 168, 2004, p 309-320.

[MSA 98] M'Saoubi R., Lebrun J.L., Changeux B., "A new method for cutting tool temperature measurement using CCD infrared technique: Influence of tool and coating", *Machining Science and Technology*, 2(2), 1998, pp.369-382.

[MSA 03] M'Saoubi R., Chandrasekaran H., "Investigation of the effects of tool micro-geometry and coating on tool temperature during orthogonal turning of quenched and tempered steel", *International Journal of Machine-Tools & Manufacture*. 2003. Vol 44. pp:213-224.

[MEI 00] Meiller P., Lebrun J.L., Touratier M., Ryckelynck D., 2000, "Friction law for tool/workpiece contact area in dry machining", *International Workshop on Friction and Flow Stress in Cutting and Forming*, 101-109.

[MER 44] Merchant E., 1944, "Basic mechanics of the metal-cutting process, Transaction of the ASME", *Jour. App. Mech.*, 66:168-175.

[MES 99] Meslin F., Hamann J.C., Caestecker P., "Experimental observation of material flow evolution with chip formation mechanism, modeling with A.L.E. and remeshing techniques", *Proceedings 2nd CIRP International Workshop on Modelling of Machining Operations*, Nantes, France, 1999, pp.336-357.

[MES 00] Meslin F., Hamann J.C., 2000, "Definition of constitutive equations and friction by inverse method and machining tests", *Int. Workshop on Friction and Flow Stress in Cutting and Forming*,112-137.

[MH1 78] Metals Handbook, Properties and Selection: Irons and Steels, ASM, 1978, Vol.1.

[MH3 80] Metals Handbook, Properties and Selection: Stainless Steels, Tool Materials and Special Purpose Metals, ASM, 1980, Vol.3.

[MEY 86] Meyer and Manwaring, *Metallurgical Applications of Shock-wave and High Strain Phenomena*, Marcel Dekker, 1986, 657-677.

[MEY 00] Meyer L.W., Abdel-Malek S., "Strain rate dependence of strength-differential effect in two steels", *Eurodymat 2000, J.Phys.*, IV France 10, Pr9-63/Pr-Pr9-68

[MOV 00] Movahhedy M.R., Gadala M.S., Altintas Y., "Simulation of chip formation in orthogonal metal cutting process: an ALE finite element approach", *Machining Science and Technology* 4(1) (2000) 15-42O.

[NG 99] Ng E., Aspinwall D.K., Brazil D., Monoghan J., "Modelling of temperature and forces when orthogonally machining hardened steel", *International Journal of Machine Tools & Manufacture*, 1999, Vol.39, pp.885-903.

[NG 00] Ng E., Aspinwall D.K., "Hard part machining AISI H13 (≈50HRc) using AMBORITE AMB90: A finite element modeling approach", *Industrial Diamond Review* 4/2000, pp.305-310.

[NG 02a] Ng E., El-Wardany T.I., Elbestawi M.A., "Influence of flank wear length on residual stress and workpiece quality in machined surfaces", *Journal of Mechanical Production Systems Engineering*, 2002.

[NG 02b] Ng E., El-Wardany T.I., Dumitrescu M., Elbestawi M.A., "Physics based simulation of high speed machining", *Proceedings of the 5th CIRP International Workshop on Modeling of Machining Operations*, West Lafayette, IN, USA, Mayo 2002, pp.1-19.

[OKU 71] Okushima K., Kakino Y, "Residual stress produced by metal cutting", *Annals of the CIRP,* Vol.20/1/1971, pp.13-14.

[OLS 89] Olsson M., 1989, "Simulation of cutting tool wear by a modified pin-on disc test", *Int. J. Mach. Tools Manufact.*, 38/1-2:113-130.

[OXL 89] Oxley P.L.B., *The Mechanics of Machining: An Analytical Approach to Assessing Machinability*, Ellis Horwood Limited, 1989.

[ÖZE 06] Özel T., "Influence of friction models on finite element simulations of machining", *International Journal of Machine Tools and Manufacture*, Vol. 46, Issue 5, 2006, pp. 518-530.

[ÖZE 08] Özel T., Karpat Y., Srivastava A., "Hard turning with variable micro-geometry PcBN tools", *Annals of the CIRP*, 2008, 57/1, pp. 73-76.

[PAN 96] Pantale O., Rakotomalala R., Touratier M., Hakem N., 1996, "A three dimensional Numerical Model of orthogonal and oblique metal cutting processes", *Engineering Systems Design and Analysis*, ASME-PD, 75:199-205.

[PIE 05] Piendl S., Aurich J. C., Steinicke M., "3D finite-element simulation of chip formation in turning", *5th CIRP Int. WMMO*, 2005, pp.225-233.

[PUJ 07] Pujana J., Arrazola P.J., M'Saoubi R., Chandrasekaran H., 2007, "Analysis of the inverse identification of constitutive equations applied in orthogonal cutting process", *International Journal of Machine Tools and Manufacture*, 47, 2153–2161

[RAT 05] Ratchev S., Liu S., Becker A.A., "Error compensation strategy in milling flexible thin-wall parts", *Journal of Materials Processing Technology*, v 162-163, 2005, p 673-681

[REC 64] Recht R.F., "Catastrophic thermoplastic shear", *Journal of Applied Mechanics*, June 1964, Vol.86, pp.189-193.

[REC 05] Rech J., Yen Y.C., Schaff M.J., Hamdi H., Altan T., Bouzakis K.D., "Influence of cutting edge radius on the wear resistance", *Wear* 259 (2005) 1168-1176L.

[ROS 91] Rosseto S., Koch U., "An investigation of temperature distribution on tool flank surface", *Annals of the CIRP*, 1971, Vol.XVIV, pp.551-557.

[SEK 93] Sekhon G.S., Chenot J.L., "Numerical simulation of continuous chip formation during non-steady orthogonal cutting", *Engineering Computations*, 1993, Vol.10, pp.31-48.

[SHA 97] Shaw M.C., *Metal Cutting Principles*, Clarendon Press, Oxford. 1997.

[SAR 04] Sartkulvanich P., Koppka F., Altan T., "Determination of Flow Stress for Metal Cutting Simulation-a Progress Report", *Journal of Materials Processing Technology*, 2004, Vol. 146, pp.61-71

[SHA 00] Shatla M., Altan T., 2000, "Flow stress measurement and applications in metal cutting". *Inter. Workshop on Friction and Flow Stress in Cutting and Forming*, Paris, France, pp.127-134.

[SHA 01] Shatla M., Kerk C., Altan T., "Process modeling in machining. Part I: determination of flow stress data", *International Journal of Machine Tools & Manufacture*, 2001, Vol. 41, pp. 1511-1534.

[SHI 74] Shirakashi T., Usui E., *Simulation Analysis of Orthogonal Metal Cutting Mechanism*, Tokyo Institute of Technology, 1974, pp.535-540.

[SHI 83] Shirakashi T., Maekawa K., Usui E., "Flow stress of low carbon steel at high temperature and strain rate (Part 1) – property of incremental strain method in impact compression test with rapid heating and cooling systems", *Bull. Japan Soc. Prec. Engg.*, 1983, Vol.17, No.3.

[SHI 03] Shirakashi T., "FEM simulation analysis on ductile mode glass machining process", *Proceedings of the 6th International ESAFORM Conference on Material Forming*, Salerno, Italy, 2003, pp.539-542.

[SIM 07] Simoneau A., Ng E., Elbestawi M.A., 2007, "Modeling the effects of microstructure in metal cutting", *International Journal of Machine Tools and Manufacture*, Volume 47, Issue 2, February 2007, Pages 368-375.

[SÖH 01] Söhner J., Weule H., Biessinger F., Schulze V., Vöhringer O., "Examinations and 3D-simulations of HSC face milling process", *Proceedings of the 4th CIRP International Workshop Modelling of Machining Operations*, Delft, The Netherlands, August 2001, pp.111-116.

[SPA 71] Spaans C., The fundamentals of three-dimensional chip curl, chip breaking and chip control, Thesis, Rotterdam, The Netherlands, 1971.

[STE 93] Stephenson D.A. "Tool-work thermocouple temperature measurements - theory and implementation issues", *Journal of Engineering for Industry*, November 1993, Vol.115, pp.432-437.

[STR 85] Strenkowski J.S., Carroll J.T., "A finite element model of orthogonal metal cutting", *Journal of Engineering for Industry*, November 1985, Vol.107, pp.349-354.

[STR 90] Strenkowski J.S., Moon K., "Finite element prediction of chip geometry and tool/workpiece temperature distributions in orthogonal metal cutting", *Journal of Engineering for Industry* 112 (1990) 313-318.

[STR 97] Strenkowski J.S., Althavale S.M., "A partially constrained eulerian orthogonal cutting model for chip control tools", *Journal of Manufacturing Science and Engineering*, November 1997, Vol.119, pp.681-688

[TAY 74] Tay A.O., Stevenson M.G., Davis G.D.V. and Oxley P.L.B., "Using the finite element method to determine temperature distributions in orthogonal machining", *Proceedings of the Institution of Mechanical Engineers*,1974,Vol.188, pp.627-638.

[TAY 76] Tay A.O., Stevenson M.G., Davis G.D.V. and Oxley P.L.B., "A numerical method for calculating temperature distributions in machining, from force and shear angle measurements", *Int. J. Mach. Tool Des. Res*. 1976, Vol.16, pp.335-349.

[TOU 02] Tounsi N., Vincenti J., Otho A., Elbestawi M.A., "From the basic mechanics of orthogonal metal cutting toward the identification of the constitutive equation", *International Journal of Machine Tools & Manufacture*, 2002, Vol.42, 1373-1383.

[TWS 00] Third Wave AdvantEdgeTM, Theoretical Manual, v3.5, Third Wave Systems, 2000.

[TWS 03] Third Wave Systems, AdvantEdgeTM, User's Manual, v4.1, v4.2, v4.3, 2003.

[TRE 91] Trent E.M., 1991, *Metal Cutting*, Butterworth-Heinemann, London, 1991.

[UHL 07] Uhlmann E., Graf von der Schulenburg M., Zettier R., "Finite element modeling and cutting simulation of Inconel 718", *Annals of the CIRP*, 56/1.61-64.

[USU 73] Usui E., Shirakashi T., "Friction characteristics on tool face in metal machining", *JSPE*, 39(9) (1973) 966-971

[UED 93] Ueda K., Manabe K., "Rigid-plastic FEM analysis of three-dimensional deformation field in chip formation process", *Annals of the CIRP*, Vol. 42/1/1993, pp.35-38.

[UED 01] Ueda T., Hosokawa A., Oda K., Yamada K., "Temperature on flank face of cutting tool in high speed milling", *Annals of the CIRP*, Vol.50/1/2001. pp.37-40.

[VAN 98] Van Luttervelt C.A., Childs T.H.C., Jawahir I.S., Klocke F., Venuvinod P.K. "Keynote papers", *Annals of the CIRP*, Vol.47/2/1998, pp.587-626.

[VEN 93] Venkatesch V.C., "A study of chip surface characteristics during machining of steel", *Annals of the CIRP*, Vol.42/1/1993, pp.631-636.

[WAR 77] Warnecke G., "A new method of visualizing the cutting process", *Society of Manufacturing Engineers*, 1977, pp.229-236.

[WAR 02] Warnecke G., Oh J.-D., 2002, "A new thermo-viscoplastic material model for finite-element-analysis of the chip formation process", *Annals of the CIRP*, 51/1:79-82.

[WRI 78] Wright P.K., "Correlation of tempering effects with temperature distribution in steel cutting tools", *Journal of Engineering for Industry*, May 1978, Vol.100, pp.131-136.

[YEO 00] Yeo S.H., Ong S.H., "Assessment of the thermal effects on chip surfaces", *Journal of Materials Processing Technology*, 2000, Vol.98, pp.317-321.

[ZER 87] Zerilli F.J., Armstrong R.W., "Dislocation-mechanics-based constitutive relations for material dynamics calculations", *Journal Applied Physics*, 1987, 61 (5), p.1816-1825.

[ZEM 07] Zemzemi F., Rech J., Ben Salem W., Dogui A., Kapsa P., 2007. "Development of a friction model for the tool-chip-workpiece interface during dry machining of AISI4142 steel with TiN coated carbide cutting tools". *International Journal of Machining and Machinability of Materials*, Vol. 2, Nos. 3/4.pp:361-377

Chapter 4

Computational Modeling of Machining Systems

This chapter deals with applications of artificial intelligence techniques for modeling machining systems and processes. Other non-conventional computational methods are also included. It is mainly focused on describing the most noteworthy reported approaches, remarking its advantages and drawbacks.

4.1. Introduction

4.1.1. *Relevance of the modeling of machining systems*

In machining, as well as in any other manufacturing or industrial process, modeling of its physical and technical parameters is a key issue, not only for understanding the nature of involved phenomena, but also for planning, optimizing and controlling the process itself.

Explicit mathematical models are required for studying and understanding the physical principles of metal cutting. In this sense, not only the analytical models but also the empirical models are very useful. These allow us to discover the relationship between variables involved in the

---

Chapter written by Ramón QUIZA and J. Paulo DAVIM.

process, glimpsing the possible physical phenomena and validating their analytical description.

On the other hand, models are indispensable for planning and optimizing cutting processes, since they relate the so-called decision variables with the optimization targets and with the problem constraints. The accuracy of these models has a heavy influence on the outcomes of the optimization process.

Moreover, automatic monitoring and controlling of the machining processes play an important role in modern intelligent manufacturing systems. In order to implement really useful and practical monitoring and control systems, there is a need to develop models to relate the monitored (controlled) variable with other parameters that can be easily measured.

Because of its importance, a great amount of research has been done in this field, however, there is not yet a universally accepted technique for accurate modeling of the cutting process. Due to the complex nature of machining processes, this is a very ambitious objective, and, it will probably pass a lot of time before it is reached.

4.1.2. *Brief historical note*

Metal cutting models may be classified as analytical and empirical. In the first models, functional relationships between involved variables are determined from physical laws, according to the nature of the studied phenomena. On the contrary, in the second models, relationships are only extracted from experimental data without any theoretical knowledge.

Due to the nature of the cutting processes, which involve plastic deformations, heat transfer, diffusion, chemical reactions and other complex phenomena, in machining, empirical models have been more successful than analytical ones.

In 1907, Taylor proposed the first known model for a cutting process, relating tool life T and cutting speed v to the expression:

$$vT^n = C \qquad [4.1]$$

where n and C are constants depending on tool and work material, tool geometry and cutting conditions other than cutting speed (feed, depth of cut, fluids, etc.). Values of C and n are calculated from experimental data by using regression.

Later, Taylor's equation was rearranged and generalized to include other cutting parameters, like feed f and depth of cut a:

$$T = \frac{C'}{v^{n_1} f^{n_2} a^{n_3}} \quad [4.2]$$

Empirical models relating other variables, such as cutting forces or surface roughness and cutting parameters, were also obtained in the form of regression equations. Occasionally, other variables such as hardness or tool geometry were included in these equations.

Due to the above-mentioned difficulties, the development of analytical models for cutting processes was very slow. They were firstly focused on explaining the chip formation process. Pioneers in this subject were Tresca and Mallock in the last quarter of the 19th century. Instead they do not propose mathematical models, they identify important aspects in chip formation, such as plasticity and friction in the chip-tool interface.

The first analytical model for a cutting process was actually proposed by Merchant in 1945, describing the chip formation in orthogonal cutting. Works by Lee and Shaffer in 1951, and Oxley and Palmer in 1959 [CHI 00] were also milestones in analytical modeling.

In spite of their unquestionable theoretical relevance, these analytical models did not allow calculation of the cutting variables accurately enough to be used in practical applications. Therefore, empirical models were used for decades in workshop environments, and are still widely used today.

However, although statistical regression may work properly for metal cutting modeling, this technique may not describe the underlying non-linear complex relationship between process variables [MUK 06] precisely. Due to the high complexity and non-linearity of cutting phenomena, especially by using modern tool materials and high cutting speeds, prediction of cutting variables by means of traditional statistical regressions, which use linear or linearized models, has hard limitations.

The massive introduction of microcomputers in the late 1970s and early 1980s, produced an important change in metal cutting modeling, not only in analytical models but also in the empirical ones, by supplying a high computing power for researchers and engineers.

In analytical modeling, the development of powerful numerical techniques to solve differential equations must be noted, especially the finite element method. Early finite element analyses in cutting processes were performed in the 1980s, by Usui and Shirakashi (1982) and Iwata and coworkers (1984). From this date to the present, a lot of work has been done on applying the finite element method to solve different cutting problems, including effects of geometric and process parameters, thermal aspects, residual stresses, dynamic analysis and control of machine tools, wear and failure and chip formation mechanism [MAC 03].

On the other hand, in empirical modeling, the most significant change was the introduction of artificial intelligence techniques. These tools may detect and establish complex functional relationships, from a large amount of noisy data, overcoming the drawbacks of regressions. In 1987 Rangwala and Dornfeld proposed neural networks for detection of tool wear states and they have been widely used in machining modeling [SIC 02] later on. Not only neural networks, but also other artificial intelligence techniques, such as fuzzy logic [CHI 07, YUM 06], support vector machines [DON 06, THA 06], neuro-fuzzy systems [PAN 07, SHA 07, FU 06] and hidden Markov models [KAS 06, LIA 06], have been widely applied in this branch.

Advances in intelligent techniques have allowed us to obtain significant successes in some traditionally difficult problems, such as tool wear prediction and tool fault detection [SIC 02]. However, the final solution for cutting processes modeling is still far away and much work must be done in this area.

4.1.3. *Off-line versus on-line modeling*

It is commonly accepted that there are two ways for carrying out the modeling of variables involved in the cutting process. These ways are known as off-line and on-line modeling.

The first one includes the relationships between stationary parameters, i.e., variables that are not affected by the time, or where the time influence is neglected. On the contrary, in on-line modeling, the relationships between transient parameters are identified, i.e., those which change with time and, therefore, must be measured during the cutting process.

Off-line modeling was previously developed. It mainly aims to establish the functional relationships between some important physical or technological variables, such as cutting temperature, tool life and cutting force, and the specific cutting conditions.

These models not only aid the understanding of the principles ruling the cutting processes, but also enable the planning of rational and efficient processes. They play a crucial role in the optimization of cutting processes [MUK 06].

An off-line model can be expressed as:

$$y = H(\mathbf{x}) \qquad [4.3]$$

where y is the modeled variable, $\mathbf{x}$ is the vector of cutting conditions (for example, in turning, $\mathbf{x} = [v, f, a]^T$, where v is the cutting speed, f the feed and a the depth of cut) and H is a non-linear operator that depends upon the modeling technique (analytical, statistic or artificial intelligence-based). Amongst the cutting conditions may be included not only the cutting parameters but also other variables such as working and tool material properties, tool geometry, machine-tool features, cutting fluid, etc.

On-line modeling arises with the automation of a cutting process. It frequently tries to find the relationship between features extracted from measured signals (components of the cutting force, vibrations, noise, etc.) and cutting conditions on the one hand, and tool wear or tool fracture on the other hand [SIC 02]. This modeling is also called indirect monitoring. An on-line model takes the general form:

$$y(t) = H[\mathbf{x}, \mathbf{z}(t)], \qquad [4.4]$$

where $\mathbf{x}$ is the vector of cutting conditions and $\mathbf{z}$ is the vector of on-line measured parameters.

Due to the drawbacks of current analytical models, operator *H*, in off-line as well as in on-line models, is most often determined empirically, from experimental machining data [CHI 00].

4.2. Computational tools for modeling of machining systems

4.2.1. *Preliminary considerations*

Besides statistical regressions, the most employed computational techniques in cutting process modeling are based on artificial intelligence. They are, mainly, artificial neural networks, fuzzy logic and neuro-fuzzy systems. These techniques are included in a branch also known as soft computing.

Furthermore, some non-conventional techniques, such as vector support machines, hidden Markov models and Bayesian probability theory, have also been used in machining modeling, so they are included in this study. In the following sections all these tools will be briefly explained.

4.2.2. *Artificial neural networks*

Artificial neural networks are connectionist structures attempting to simulate the functioning of a human brain. They utilize interconnected mathematical nodes, so-called neurons, to form a network that can model complex functional relationships. A neuron consists of a linear activator followed by a non-linear inhibiting function (see Figure 4.1). The linear activation function yields the sums of weighted inputs, while the non-linear inhibiting function attempts to arrest the signal level of the sum.

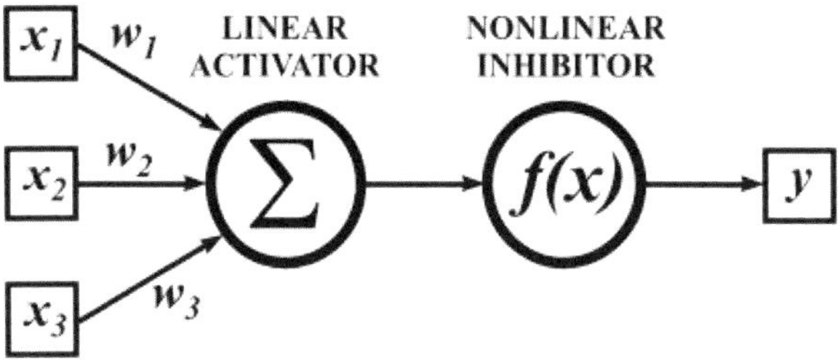

Figure 4.1. *Logical scheme of a neuron*

Therefore, the neuron output may be calculated by:

$$y = f\left(\sum_i w_i x_i\right) \quad [4.5]$$

where y is the output, x_i are the inputs, w_i are the weights and $f()$ is the inhibiting function, sometimes called the activation function. Usually, an independent term, so-called bias b, is added to the weighted sum:

$$y = f\left(b + \sum_i w_i x_i\right) \quad [4.6]$$

A neural network is a set of these neurons arranged in a certain topology. Besides any neurological approach, neural networks can be understood as a class of general, flexible, non-linear regression models.

A neural network can be conventionally represented as a directed graph (see Figure 4.2), where neurons are arranged into layers. Depending upon their position, the layers are called input layer, hidden layer and output layer. A network may contain several hidden layers.

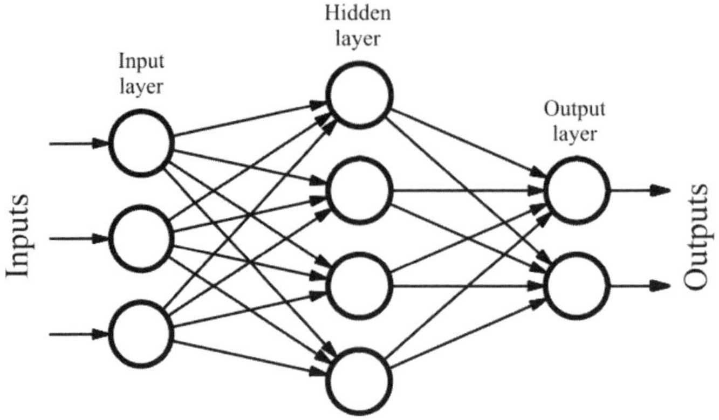

Figure 4.2. *Graph of a feed-forward neural network*

If a network graph does not include cycles, i.e., if only forward connections exist, then it is called a feed-forward or non-recurrent neural network. In this group are included multilayer perceptrons (MLP), radial basis function networks (RBF) and self-organizing maps (SOM).

On the contrary, if there is any cycle or feedback into the network graph, it is called a recurrent neural network (see Figure 4.3). Elman networks and Hopfield networks are typical samples of recurrence.

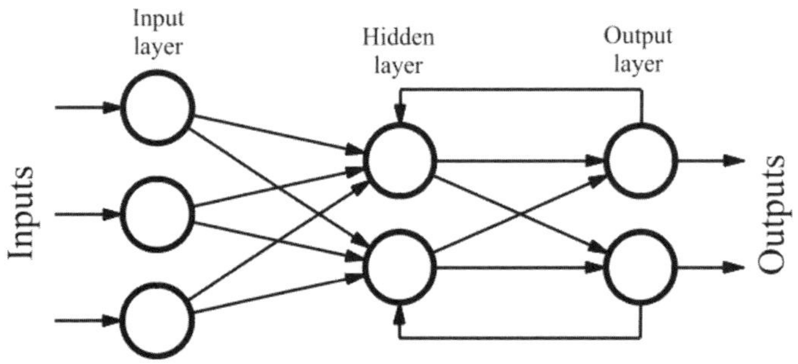

Figure 4.3. *Graph of a recurrent neural network*

A neural network is actually a function constructed from compositions of weighted sums of bounded monotone functions. The following is an example of a typical two-layer[1] non-recurrent neural network:

$$f_i(x) = f^{out}\left[b_i^{out} + \sum_{j=1}^{n} w_{ij}^{out} f\left(b_j + \sum_{k=1}^{m} w_{jk}^{hid} x_k\right)\right] \qquad i = 1 \ldots p \qquad [4.7]$$

where the functions f^{out} and f^{hid} are the activation functions (superscripts *out* and *hid*, indicate the output layer and hidden layer, respectively); x_i are the inputs; w_{ij} are the weights; b_i are the biases; and m, n and p are the number of neurons in the input, hidden and output layers. Weights and biases are free parameters, which must be determined from the data set of inputs and outputs. The process of adjusting weights and biases is known as training, and a used data set is called a training set.

The process of training a neural network can be broadly classified into two typical categories:

– supervised learning: requires a trainer that submits both the input and the target patterns for the objects to become recognized; includes back-propagation for MLP, recurrent neural networks (RNN), time delay neural networks (TDNN) and radial basis function networks (RBF);

– unsupervised learning: is required in many recognition problems, where the target pattern is unknown; includes adaptive resonance theory (ART) and SOM, also known as Kohonen feature maps (KFM).

There are a lot of training methods for different network types. The most popular is back-propagation, which is applied to multilayer perceptrons. It is a gradient-descent method that minimizes the mean-square error between the network outputs and the set of training data.

Non-linear function approximation is one of the most important applications of multi-layer neural networks. It has been tested whether a two-layer neural network can approximate any continuous function

1 When counting the layers of a network, the input layer is often not considered. Hence, a two-layer network actually has three layers: input, hidden and output.

arbitrarily well, provided that it has a sufficient number of neurons in the first layer. This is the so-called universal approximation property.

4.2.3. *Fuzzy logic*

Fuzzy logic is a generalization of conventional (Boolean) logic that deals with fuzzy sets and logical connectives for modeling the human-like reasoning problems of the real world. It was proposed by Zadeh in the 1960s. In opposition to traditional logic, where the prepositions can only assume two values, either "false" (0) or "true" (1), fuzzy logic can deal with several degrees of truth.

A fuzzy set is a subset of elements, each having an associated value, from the interval [0, 1] which defines its membership to a certain set. These values are also known as the degree of truth, and their distribution is called the membership function.

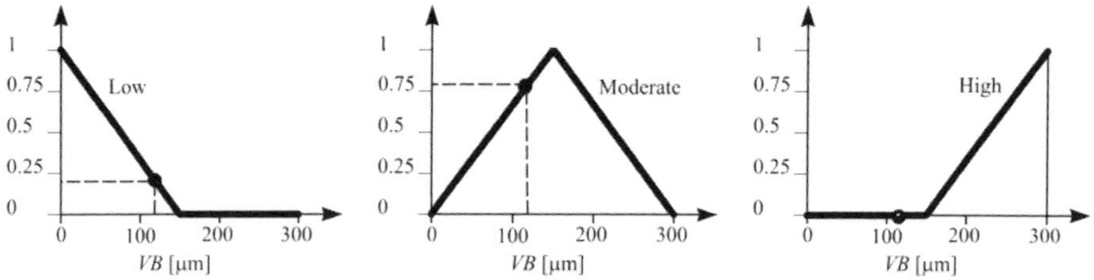

Figure 4.4. *Samples of membership functions*

For example, in Figure 4.4 membership functions are shown for three fuzzy subsets of tool wear level, *VB*, called low, moderate and high. Therefore for a wear VB = 120 μm, the following statements have the indicated degree of truth:

$$\left.\begin{aligned} &VB \text{ is } \text{``}Low\text{''} = 0.20 \\ &VB \text{ is } \text{``}Moderate\text{''} = 0.80 \\ &VB \text{ is } \text{``}High\text{''} = 0.20 \end{aligned}\right\} \qquad [4.8]$$

A general fuzzy inference system consists of three parts. A crisp input is firstly fuzzified by expressing the input variables in the form of fuzzy membership values based on various membership functions. Then, a fuzzy rule base processes it to obtain a fuzzy output. Finally, this fuzzy output is defuzzified to give a crisp outcome.

A fuzzy rule base consists of a collection of m fuzzy if-then rules in the following form:

$$\text{if } (x_1 \text{ is } X_1^i) \text{ and} \ldots \text{and } (x_n \text{ is } X_n^i) \text{ then } (y \text{ is } Y^i), \quad i = 1 \ldots m, \qquad [4.9]$$

where A_j^i and Y^i are fuzzy sets. For example:

$$\left.\begin{array}{r} \text{if } (force \text{ is ``}high\text{''}) \text{ and } (time \text{ is ``}high\text{''}) \text{ then } (VB \text{ is ``}high\text{''}) \\ \text{if } (force \text{ is ``}moderate\text{''}) \text{ and } (time \text{ is ``}high\text{''}) \text{ then } (VB \text{ is ``}moderate\text{''}) \\ \text{if } (force \text{ is ``}low\text{''}) \text{ and } (time \text{ is ``}low\text{''}) \text{ then } (VB \text{ is ``}moderate\text{''}) \\ (\ldots) \end{array}\right\}$$

Usually fuzzy rules are extracted from experimental data or from human experience, so they are empirical.

Depending on the type of reasoning and fuzzy inference rules employed, fuzzy inference systems can be classified into three types:

– Tsukamto inference systems: the overall output is the weighted average of each rule’s crisp output induced by the rule’s firing strength and output of the membership function;

– Mandami inference systems: the overall output is obtained by applying the “max” operation to the qualified fuzzy outputs; then the final fuzzy outcome is crisped by an appropriated scheme (centroid of the area, bisector of the area, etc.);

– Sugeno inference systems: use Tagaki and Sugeno’s fuzzy if-then rules, where the output of each rule is a linear combination of input variables plus a constant term, and the overall output is the weighted average of all of rules’ output.

4.2.4. *Neuro-fuzzy systems*

Combining neural networks and fuzzy logic may result in a more powerful system due to its ability to joint verbal power of fuzzy systems with the numeric power of neural networks. While neural networks support the system to have the capability of training from empirical data, fuzzy logic provides reasoning for the training process and generates a rule bank for control or classification purposes.

Although there are many possible combinations of the two systems, there are four basic combinations that have been successfully applied (see Figure 4.5) [LI 00]. The first combination (Figure 4.5a) uses the neural network (NN) to optimize the parameters of the fuzzy system (FS) by minimizing the gap between the output of the fuzzy system and the given target. In the second combination (Figure 4.5b) the output of a fuzzy system is corrected by the output of a neural network to increase the precision the final system output. The third and fourth combinations (Figure 4.5 c-d) are cascade assemblies of a neural network and a fuzzy system.

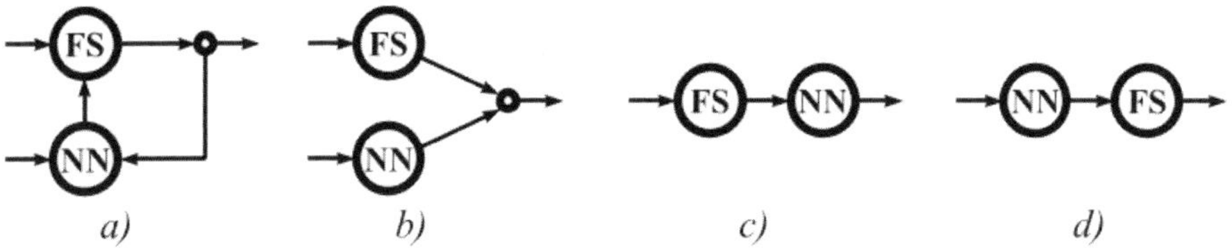

Figure 4.5. *Possible combinations of neural networks and fuzzy systems [LI 00]*

In neuro-fuzzy inference systems, unlike fuzzy systems based on Mamdani or Sugeno-type inference systems, fuzzy parameter identification is automated with the use of neural networks.

An adaptive neuro-fuzzy inference system (ANFIS) is a multilayer feed-forward network using neural network learning algorithms and fuzzy reasoning to model relations between given input/output data sets.

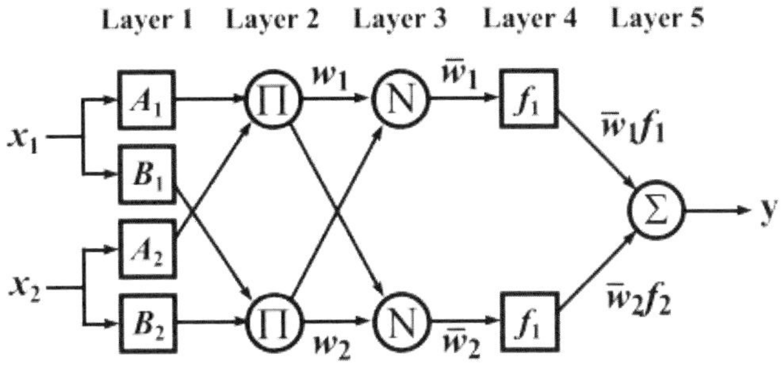

Figure 4.6. *ANFIS architecture*

As shown in Figure 4.6, some of the nodes are circular (with fixed transfer function) and the others are square (with adaptive parameters). x_1 and x_2 represent inputs and A_1, B_1, A_2, B_2 represent fuzzy sets of input variables. Layer 1 is the fuzzify layer which provides transition between numerical and fuzzy values determining membership degrees. Layer 2 is the product layer which multiplies its inputs, in order to generate one output combining both rules. Layer 3 is the normalization layer that calculates the firing strength of each rule. Layer 4 is a defuzzification layer, which calculates fuzzy rule results by using a linear function such as:

$$f = px_1 + qx_2 + r \quad [4.10]$$

where p, q and r are the linear parameters of Sugeno inference system and they can change in order to learn. Finally, layer 5 is the total output layer, which realizes mathematical calculation, giving the ANFIS output.

4.2.5. *Other non-conventional techniques*

4.2.5.1. *Support vector machines*

Support vector machines were initially developed by Vapnik in 1995 for classification problems with separable data, and it was later improved to handle non- separable data. Support vector machines can take advantage of prior knowledge for constructing a hyperplane to be used as the decision surface, so that the margin of the separation between different classes is maximized.

Support vector machines exploit the idea of mapping input data into a high-dimensional space. In addition, the support vector regression methods have many other advantages, including a global minimum solution as the minimization of a convex programming problem, relatively fast training speed, and sparseness in solution representation.

Although support vector machines were initially conceived to separate only two classes (binary classification) there are some strategies allowing carry out multi-class classification tasks, such as one-against-all (OAA), one-against-one (OAO) and direct acyclic graph (DAG).

4.2.5.2. *Hidden Markov models*

Hidden Markov models are stochastic models of process that have piecewise stationary regions, where the time evolution of the non-stationary behavior can be characterized in terms of an unobserved discrete Markov chain. A hidden Markov model is specified by the set of discrete states, initial state probabilities, transition probabilities between states and a set of state-dependent observation distributions. Standard algorithms exist for efficient searches for the maximum likelihood state sequence and for parameter estimation.

Hidden Markov models are divided into continuous and discrete models according to their probability density description. Discrete hidden Markov models can represent any distribution as no assumptions are made about the underlying distribution of the observed symbols. Hidden Markov models are useful, in general, for problems where there are temporal dynamics.

4.2.5.3. *Bayesian probability theory*

Bayesian probability theory provides a unifying framework for data modeling, which offers several benefits, such as solving the over-fitting problem and handling uncertainty in a natural manner. In the Bayesian approach, the uncertainty about model parameters is represented by a probability distribution. Principles or Bayesian probability theory have been successfully applied to neural networks and support vector machines.

4.2.5.4. *Group method of data handling*

Group method of data handling (GMDH) is a family of inductive algorithms for mathematical modeling of multi-parametric datasets that features fully-automatic structural and parametric optimization of models.

GMDH include abductive networks, polynomial neural networks and other related techniques.

4.3. Modeling for monitoring and control

4.3.1. *Tool wear monitoring*

4.3.1.1. *Relevance of tool wear modeling*

Tool monitoring systems can significantly improve the performance of machining processes by providing information on some important aspects, such as collisions, tool breakage and tool wear [SIC 02].

Tool wear modeling deals with progressive mass loss by the cutter during the machining process. This topic is a key issue in automation and optimization of machining processes, so it has attracted a great amount of research. Due to its capital importance and the wide application of intelligent techniques, this topic will be carefully reviewed.

Only considering the most recent years, many approaches have been proposed for tool wear monitoring in turning [CHE 07, SHA 07, JEM 06, RAO 06, SIL 06, THA 06], milling [LEE 07, DON 06, FU 06, KAS 06, WAN 06], drilling [PAN 07, PAT 07, SIN 06, YUM 06] and grinding [LIA 06]. However, a universal, reliable and flexible tool monitoring system is a very ambitious goal and has yet to be achieved [CHE 07, KAS 06].

Tool wear monitoring can be carried out either out-of-process or in-process. In out-of-process (or in-cycle) monitoring, machining has to be stopped to measure the wear level by using a microscope or another measuring device. On the contrary, in-process tool wear monitoring can take place during the machining process, so it leads to higher machining efficiency.

All different tool wear monitoring strategies can be broadly divided into two major classes: direct and indirect methods. Direct methods involve the measurement of tool wear directly by using mechanical, optical, electrical or spectroscopic transductors. On the other hand, indirect methods involve the measurement of parameters that can be easily measured and related to tool wear. The most used parameters in tool wear indirect measurement are cutting forces, acoustic emission, vibrations, cutting temperature, current signals and surface roughness. Although direct methods measure tool wear directly, and therefore most accurately, their implementation is costly and cannot be applied on-line. Thus, indirect methods are more popular. By and large, direct monitoring systems are out-of-process while indirect monitoring systems are in-process.

There are some important aspects in tool wear monitoring that must be carefully reviewed because they reflect the current stage of this field and its future directions.

4.3.1.2. *Wear parameters*

As tool wear is a vague concept it is necessary to use well defined parameters to reflect the tool state. The most commonly used parameter is the width of wear land, due to its more direct influence on the dimensional accuracy of the product. In measuring flank wear it used either the average value or the maximum value. Many works do not explicitly establish if they use average or maximum value, although notation *VB* seems to indicate, in most of the cases, that the average flank wear (in zone "B", as established in standards ISO 3685: 1993 or ASME B94.55M: 1985) is used. In several papers there is no indication of which parameter was used to describe wear.

An interesting approach is given by Jemielniak and Bombinski [JEM 06] who propose considering the used up portion of the tool life to evaluate the tool condition. It is an intuitive concept and it is much more appropriate for using in a workshop floor environment. However, it assumes that tool wear grows linearly with time, which is not generally true.

4.3.1.3. *Objectives of the tool wear monitoring*

From the point of view of the objective of tool wear monitoring, it can be implemented either for continuous estimation of tool wear or for its classification in several well-defined classes. Continuous estimation corresponds more closely to the physical nature of the tool wear process. Furthermore, it is more versatile and can not only be used to implement an efficient tool change policy, but also to adjust the tool position during the cutting process in order to compensate the tool length decrease due to wear. Continuous estimation of tool wear is used in a lot of works.

On the other hand, wear classification is often simpler to implement and gives a definite criterion like "sharp" or "worn". The number of used classes is noteworthy [SIC 02]. Few authors prefer only two classes; most of them actually use three.

4.3.1.4. *Measured signals*

An essential issue in tool wear monitoring is the selection of signals to be measured and used as inputs of the tool wear model, i.e., the components of the $\mathbf{z}(t)$ vector in the equation [4.4]. As can be noted, this vector depends upon time, so it must be measured every so often.

Cutting forces are often used to predict tool wear because they are easy to measure [SHA 07]. Furthermore, they have a clear phenomenological relationship with tool wear: cutting forces are affected by the tool geometry, so, as the tool becomes worn, its geometry changes impacting over cutting forces. However, there is no agreement in which cutting force component has a more closed relationship with tool wear.

In turning, most authors use the feed (axial) component while others prefer the main (tangential) or radial components. Sometimes, several components are simultaneously used, either by combining their signals or by computing their ratio. In milling, the transversal component is often used. In drilling, the thrust (feed) component of the force is measured for all reviewed cases. Also, torque is recognized in several cases as a useful signal.

The main drawback of using cutting forces is that the implementation of dynamometric measure systems on the workshop floor is restricted because of its high cost, negative impact on machining system rigidity, requirement of a wiring harness and extra space for installation.

Acoustic emission is another preferred signal related to tool wear, not only in turning but also in milling and grinding. There are several sources of acoustic emission during metal cutting process, such as plastic deformations in the work piece and chip, frictional contact between the tool flank face and the work piece and between the tool rank face and the chip, collisions between chip and tool, chip breakage and tool fracture [LI 02]. The major advantage in the use of acoustic emission in tool wear monitoring is that the frequency range of the acoustic emission signal is much higher than that of the machine vibrations and environmental noises [CHE 07]. Perhaps its major disadvantage is that there are very few microphones robust enough to operate in a practical workshop environment, and their positioning is not a simple task.

Another meaningful signal, frequently used to monitor tool wear, is cutting tool vibration, which is produced by the rubbing action at the work piece-tool flank interface, the formation of a built-up edge, and waviness of the work surface [SHA 07]. The fundamental frequency of tool vibrations is the resonant frequency of the system excited by the friction at the cutting edge. Acceleration is the signal measured most often for characterizing tool vibration [LEE 07]. The main shortcoming in the use of vibration signals is their susceptibility to contamination and their dependency on vibration characteristics on the machine-fixture-tool-work piece combination. Using vibration is reported mainly in turning where these signals are meaningful and easy to measure.

Current or power signals have been successfully used in tool wear monitoring [LEE 07, FU 06], either from spindle motor or from feed motor. Theses parameters are closely related to cutting torque and, therefore, with cutting force. Perhaps, the main advantage of using current and power signals is its simple hardware implementation that does not interfere with the process and does not require high cost devices, however it is not as sensitive to tool wear and cutting force as acoustic emission. There are reported uses of current or power signals in turning, milling and drilling.

The use of temperature has been suggested too as parameter for monitoring tool wear. This is based on the fact that wear changes the tool geometry and its ability to cut [LIA 04b]. To measure the temperature in the tool tip, either natural thermocouple, formed between the tool and the work piece, or sensors embedded or attached to the tool can be used.

Surface quality is also used to categorize the cutting tool condition. The tool operates directly on the work piece and the machined surface carries valuable information about the machining process. An attractive approach is presented by Kassim *et al.* [KAS 06] who use an image of a machined surface inferring wear from its texture.

In many cases, multiple signals, rather than a single one, are measured simultaneously. This approach is commonly known as sensor fusion and claims to produce better results [LIA 04b] even though this point of view is not unanimously accepted [SIC 02].

4.3.1.5. *Levels of tool condition monitoring systems*

A generic scheme for tool condition monitoring consists of several steps or levels. Sick [SIC 02] states the following five levels:

– analog preprocessing: deals with the preparation of raw data before digitization;

– digital preprocessing: modifies data before digitization, by filtering, normalizing, linearization and other techniques;

– feature extraction: condenses the remaining information in a few values which can be used as inputs of wear models;

– wear model: estimates or classifies the wear from values obtained in the prior level;

– decision making: takes the output of the wear model level as a preliminary result and combines several of these results to come to a final conclusion.

It must be pointed out that this scheme is only a general approach and most of the published research does not report the use of methods in each level. As a matter of fact, it can be noted that studies are mainly concentrated on feature extraction and wear model levels.

4.3.1.6. *Analog and digital preprocessing*

Very few works report the use of analog preprocessing, mainly low-pass filters for electrical signals, for vibrations and for force signals. Band-pass filters are used mainly for acoustic emission.

There is a similar situation in the digital preprocessing level. Some authors use clustering to remove redundant signal features or to identify natural groups of data. Lee *et al.* [LEE 07] propose a hybrid approach for force regulation. Dong and co-workers [DON 06] report the use of normalizing force signal parameters. An interesting approach is given by Rao *et al.* [RAO 06] who use the finite element method to predict the temperature in the tool tip from values measured in other points of the tool. Wang *et al.* [WAN 06] implement a preprocessor that multiplies each approximate coefficient of the wavelet transform with a weight just before passing them to the model level. In this case, the preprocessing level is located between the feature extraction and the model level, in an unusual arrangement. These weights are determined by applying a genetic algorithm.

4.3.1.7. *Feature extraction*

A very different situation is shown in the feature extraction level, which is necessarily considered in all the consulted approaches, although some of them are not very clear about its implementation. Features can be extracted considering the time domain, frequency domain and time-frequency domain.

Time domain features have the advantage that they do not require tedious calculations, therefore the editing, compiling and debugging of data acquisition programs is very quick and easy. The most often used time domain features are average, standard deviation or variance, skewness, kurtosis and root mean square (rms). However, these features are less informative about the process and are affected by the noises and disturbances of the system. In spite of this undoubted assertion, it can be noted that time domain features are the most widely reported.

Another approach is the use of singular spectrum analysis, which decomposes a given time series into an additive set of independent time series, being interpreted as consisting of a trend representing the signal mean at each instant, a set of periodic series, and a non-periodic noise.

Frequency domain features are given by methods estimating the distribution of power over the spectrum of frequencies. In this sense, fast Fourier transforms (FFTs) have been used to generate a power spectral density function. Nevertheless, the averaging natures of the FFT calculations tend to obscure the frequency content of transient and burst phenomena, so they are not suitable for analyzing transient signals [PAT 07].

On the other hand, wavelets are mathematical functions that cut up data into different frequency components and then study each component with a resolution matched to its scale [CHE 07]. Wavelet transform has a good resolution in frequency and time domain synchronously; it can extract more information in a time domain at different frequency bands [PAT 07]. It is favored over FFT because of their variable resolution. Wavelet analysis is sometimes used for tool wear estimating and classifying.

A noteworthy approach, strongly differing from all the previously mentioned ones, is shown in [KAS 06], where fractal analysis is used to extract effective parameters from images of machined surface texture.

4.3.1.8. *Model level*

Model level is the heart of a monitoring system. It extracts tool wear values from features of the measured (and probably preprocessed) signals. The most important issue in model constitution is which tool is selected to identify the relationship between dependent and independent variables.

Traditionally, statistic techniques were the most popular in tool wear modeling and they remain used today [CHE 07, LEE 07, THA 06]. However, as has been repeatedly pointed out, the complexity of cutting process limits the uses of these simple techniques and has caused the introduction of modern "intelligent" tools.

Unquestionably, the most widely used of these "intelligent" techniques are the artificial neural networks, which were briefly described in section 4.2.2. Neural networks are implemented principally for estimating tool wear. The most used paradigms are multi-layer perceptrons, radial basis function networks, self-organizing maps and adaptive resonance theory (ART2). The advantage of neural networks is their ability to perform continuous estimations. Their disadvantages are their relative complexity and also that a great deal of experience and trial-and-error work is required for successfully implementing a neural-based monitoring strategy system.

The use of neuro-fuzzy systems is widely extended in tool wear monitoring. ANFIS are proposed by Sharma *et al.* [SHA 07] and Yumak and Ertunc [YUM 06]. In this last paper the performance of ANFIS is compared with fuzzy logic's, finding that fuzzy logic offers better outcomes but requires expert knowledge. Another neuro-fuzzy approach is the use of fuzzy neural networks, such as B-spline based [FU 06] or fuzzy back-propagation [PAN 07].

Another important tool that is used in tool wear modeling is the hidden Markov model (HMM). This is implemented mostly for tool wear classification but, some approaches using HMM to estimate wear values have also been proposed. The advantages of HMMs are that they are fairly easy to initialize and implement. However, HMMs typically contain a large number of parameters and therefore need large amounts of data.

Furthermore, there are other techniques which are not as widely used but sometimes reported. In this group there are: support vector machines,

Bayesian support vector machines and Bayesian neural networks, analytic hierarchy process and abductive networks.

4.3.1.9. *Decision making*

Decision making is the most poorly included level in tool wear monitoring reported research. Only two of the consulted works indicate its use. Silva *et al.* [SIL 06] implement a fuzzy inference system to combine the outcomes from a SOM and an ART2 network. Jemielniak and Bombinski [JEM 06] use a hierarchical strategy to combine the results of a MLP and a third degree polynomial approximation.

4.3.2. *Tool breakage and fault detection*

While tool wear is an essentially slow and progressive phenomenon, tool breakage is a sudden and mostly unexpected event which requires real-time reactions [SIC 02]. Tool breakage is a major cause of unscheduled stoppage in a machining environment. Some estimates state that the amount of downtime due to cutter breakage on an average machine tool is up to 20% [REH 05].

Tool breakage and fault detection are, in many aspects, similar to tool wear monitoring and are often considered together [SIC 02], however these topics have dissimilarities in some aspects. First of all, tool breakage is implemented by classification tasks while in tool wear monitoring estimation is most often used. On the other hand, at the moment when faults occur, the output signals usually contain jump points that are often singularity points. Therefore, singularity detection has played an important role in fault monitoring [CHE 07].

Approaches have been reported in recent years which use intelligent techniques for monitoring tool failure in turning, milling and boring. All of them show positive achievements.

The most commonly measured signals for tool breakage monitoring are cutting forces, spindle torque and spindle power, although sometimes multi-sensor approaches are also presented. Preprocessing is sometimes carried out by filtering, such as in [ALA 06] where a second-order Butterworth filter is implemented; however, these are not widely applied techniques.

In order to extract features from measured signals, we not only use time domain techniques like moving-average [HSU 08], but also time-frequency domains techniques such as discrete Daubechies wavelet transform [ALA 06] or joint time-frequency signal analysis [YU 06]. Alaniz and co-workers [ALA 06] compare the Fourier transform and the wavelet transform, showing that the wavelet offers better outcomes.

For the classifying tool stage, support vector machines [HSU 08], multilayer perceptrons [ALA 06] and Elman recurrent neural networks [YU 06] are used. In [CHO 05], results are compared with those of the multiple regression. This comparison, unfortunately, is not considered in other researches.

4.3.3. *Monitoring of other parameters*

4.3.3.1. *Chatter*

Chatter or self-excited vibration in machining processes has been recognized as one of the crucial factors limiting productivity and is considered to be the most serious cause of surface quality deterioration.

In recent years, some interesting approaches have been proposed for applying soft computing to chatter detection and suppression. Liang *et al.* [LIA 04a] develop a fuzzy inference system to detect chatter in milling by measuring vibrations signals. Kong and Yu [KON 07] also use fuzzy logic to predict the limits where chatter begins in a milling process.

4.3.3.2. *Surface roughness*

Surface roughness is a widely used index of product quality and in most cases is a technical requirement for mechanical products. Therefore, on-line monitoring of obtained surface roughness is important not only for evaluating the performance of the machining process, but also for taking proper corrective actions. In specialized literature some interesting work has recently been published on on-line surface roughness monitoring using artificial intelligence tools, mainly in turning processes [ONW 08, KIR 06, SON 06].

Cutting forces and vibrations are commonly used parameters for estimating the surface roughness, due to their clear, direct relationship with this variable. Most of these approaches also use cutting parameters as inputs for the surface roughness models.

To build a relationship between roughness and measured variables, different tools are implemented, such as a hybrid of differential evolution and group method for data handling, multilayer preceptrons and radial basis function neural network and neuro-fuzzy approaches.

4.3.3.3. *Delamination of composite materials*

Delamination is the phenomenon which take places in composite materials, when cutting forces become higher than the interlaminate strength, generating fracture in the tool entry and exit. Damage caused for delamination is characterized by several parameters: delamination factor (ratio of maximum diameter of the affected zone and nominal hole diameter), push-up and push-down-damage (ratio of the area concerned by the fractured inter-laminate zone occurring in the push-up or push-down phenomena, respectively, and the perimeter of the hole).

On-line prediction of delamination factors in drilling a composite material, by using a multilayer perceptron, is carried out by Dini [DIN 03]. Four neural networks are implemented to estimate and classify push-up and push-down-damage, respectively. For push-up-damage two inputs are required: thrust force and feed rate, while the push-down-damage model needs three inputs: the two previously mentioned and also drill diameter. The neural networks have been trained to adopt a back-propagation procedure with the Levemberg-Marquardt algorithm. The early stopping technique has also been applied in order to prevent overfitting.

Another similar approach given by Enemouh and El Gizawy [ENE 03], also uses force signals for monitoring process-induced damage in drilling carbon fiber reinforced epoxy composites. The model is implemented through a multilayer perceptron. The output vectors from the neural network include delamination, damage width and hole surface roughness.

4.3.4. *Intelligent control*

Process control plays an important role in modern machining. It is generally classified as [LIA 04b]:

– Adaptive Control with Constraints (ACC): process parameters are manipulated in real-time to maintain a specific process variable, such as force or power, at a constraint value;

– Adaptive Control with Optimization (ACO): machine settings are selected to optimize a performance index, such as production time, unit cost; and

– Geometric Adaptive Control (GAC), which seeks to maximize the quality of finishing operations in the face of structural deflections and tool wear.

For different machining processes, approaches have been proposed applying soft computing techniques. The most commonly controlled variables are cutting force and surface roughness, and control input is often feed rate.

4.4. Modeling for process planning and optimization

4.4.1. *Modeling of tool wear and tool life*

Tool wear and tool life are key issues in cutting process planning and optimization. Not only does tooling cost represent an important fraction in the total operation cost, but also wear has a remarkable influence in cutting forces, vibration and work piece quality. Most of the optimization researches aim to obtain an economic equilibrium between labor cost and tooling cost in machining processes, i.e., between productivity and tool life.

In this section we considered the modeling of tool wear and tool life, for carrying out process planning and optimization. Therefore, we dealt with off-line modeling, where dependent variables are described as functions of stationary parameters, often cutting parameters. Artificial intelligence-based techniques are reportedly being used in modeling tool wear or tool life in turning mainly [QUI 08, WAN 08, BAS 07, NAT 07, OZE 07a, OZE 07b, SAR 07], but also in milling [CHI 07].

Most of the reported works in this field use neural networks to model tool wear or tool life. The most used paradigm is the multilayer perceptron. However, other network types are also used such as radial basis function network, abductive network and fully forward connected network. Sarma and Dixit [SAR 07] propose a hybrid system which combines a multilayer perceptron and a radial basis function network. Fuzzy logic is also used to model tool wear.

Most of these works use cutting parameters and time as input variables to predict tool wear. Another approach is to model tool life from cutting parameters only.

All of these works claim to obtain more accurate outcomes by using neural networks than using traditional tools, but only a few of them actually compare their outcomes with those obtained by other techniques.

4.4.2. *Modeling of surface roughness*

Several off-line surface roughness models, using soft computing techniques, have recently been proposed. As for other variables, in surface roughness the most studied machining process is turning [BAS 07, OZE 07a, OZE 07b, BAG 06]. However, works on milling [WAN 05] and drilling [SAN 06] have also been reported.

Neural networks, specifically multilayer perceptrons, are widely used in this field. Radial basis function, fuzzy adaptive networks and ANFIS have also been used.

Another important aspect is the variables used as inputs. All the above-mentioned works considered cutting parameters in their respective models. Furthermore, tool nose radius, cutting time, number of cutter teeth, work piece material, tool characteristics and the number of elaborated holes are occasionally included as input variables in surface roughness models.

4.4.3. *Modeling of cutting forces*

Modeling of cutting forces is often made difficult by the complexity of the tool/work piece geometry and cutting configuration. The large number of complex interrelated machining parameters that influence the cutting forces makes it extremely difficult to develop proper analytical models [CUS 06]. However, there are far fewer reported works on applying artificial intelligence techniques to cutting force modeling than to tool wear or surface roughness.

Sheikh-Ahmad *et al.* [SHE 07] use the specific cutting energy to predict cutting force in milling fiber reinforced polymers. To model the specific cutting energy they compare performance of a committee neural network (based on multi-layer perceptrons trained by back-propagation) and a statistical regression. Input variables are fiber orientation and uncut chip thickness. The neural model was shown to outperform the standard form of regression by smoothing the noisy data and capturing the inherent non-linearity in the experimental data. The neural model accuracy was 2.5 to 8 times better than that of the regression model.

Cus *et al.* [CUS 06] model the three components of the cutting force in a milling process by implementing a multilayer perceptron. Input variables are used: work piece data (type of material and hardness), cutting tool characteristics (diameter and type of insert), use of cutting fluid, cutting parameters (speed, feed rate and radial and axial depths of cut) and tool wear.

4.4.4. *Modeling of other parameters*

Modeling of work piece dilatation in a turning process is carried out by Segonds *et al.* [SEG 06] by using a multilayer perceptron. Feed rate, cutting speed, depth of cut, time and final diameter of the work piece are used as input variables. Outcomes of the network are tested with an independent data set, analyzing not only the magnitude of residuals but also their distribution.

In another work, Kang *et al.* [KAN 06] estimate thermal deformations in a machine tool by a hybrid system composed of an autoregressive moving-average filter and a feed-forward neural network.

Material rheological behavior is modeled by a neural network in [GIO 03]. Outputs of this model are used to supply the material properties to a finite element model of orthogonal cutting.

4.5. A case study

In order to illustrate the application of neural networks in modeling of machining systems, the following case study is presented [QUI 08]. In this sample, we analyze the process of turning hardened cold work tool steel with ceramic tools. The relationship between the cutting parameters (cutting speed, V, and feed, f) and flank wear (VB_C) is obtained using a neural network, establishing the corresponding model.

This is a typical case of off-line modeling where a dependent variable (tool wear) is described as a function of independent variables (feed, cutting speed and time) which do not change with time.

4.5.1. *Experimental set-up*

Machining experiments were performed using a CNC lathe with 18 kW spindle power and a maximum spindle speed of 4500 rpm. The levels for independent variables are given in Table 4.1. A constant depth of cut of 0.2 mm was kept.

Variable	Levels		
	Low	Medium	High
Feed [mm/rev]	0.05	0.10	0.15
Speed [m/min]	80	115	150
Time [min]	5	10	15

Table 4.1. *Levels for independent variables*

Work piece material was an AISI D2 tool steel, with the following chemical composition: 1.55% C; 0.30% Si; 0.40% Mn; 11.80% Cr; 0.80% Mo and 0.80% V. After heat treatment (quenching in a vacuum atmosphere at 1000-1040°C) an average hardness of 60 HRc was obtained.

Mixed alumina inserts with ref. CC650 (ISO code-CNGA 120408 T01020) were used to machine the tool steel with a geometry as follow: rake angle -6°, clearance angle 6°, edge major tool cutting 95° and cutting edge inclination angle -6°. An ISO DCLNL2020K12 tool holder was used.

The evaluation of the flank tool wear was made by a shop microscope with 30x magnification and 1μm resolution. The admissible wear was established according to ISO 3685 standard (1993) and measured at corner radius (VB_C).

No.	Cutting speed, *Vc* [m/min]	Feed, *f* [mm/rev]	Time, *t* [min]	Wear, *VB* [mm]
1	80	0.05	5	0.032
2	80	0.05	10	0.068
3	80	0.05	15	0.112
4	80	0.10	5	0.086
5	80	0.10	10	0.134
6	80	0.10	15	0.190
7	80	0.15	5	0.089
8	80	0.15	10	0.148
9	80	0.15	15	0.156
10	115	0.05	5	0.081
11	115	0.05	10	0.097
12	115	0.05	15	0.146
13	115	0.10	5	0.102
14	115	0.10	10	0.131
15	115	0.10	15	0.177
16	115	0.15	5	0.090
17	115	0.15	10	0.111
18	115	0.15	15	0.152
19	150	0.05	5	0.104
20	150	0.05	10	0.137
21	150	0.05	15	0.179
22	150	0.10	5	0.085
23	150	0.10	10	0.151
24	150	0.10	15	0.242
25	150	0.15	5	0.144
26	150	0.15	10	0.262
27	150	0.15	15	0.333

Table 4.2. *Experimental data [QUI 08] (© 2007, Springer Science and Business Media)*

The data obtained is shown in Table 4.2 and they are plotted in Figure 4.7. For every combination of feed and speed, as was expected, wear grows with time. However, there is a complex relation between cutting parameters and tool wear. Presumably, the combination of feed and speed has an influence on the tool wear as significantly as each single factor has.

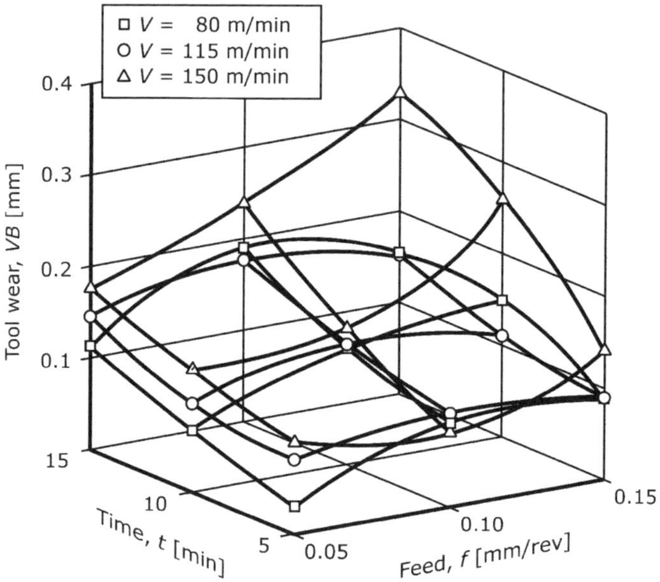

Figure 4.7. *Experimental data. Wear plotted versus time and feed for different values of cutting speed [QUI 08] (© 2007, Springer Science and Business Media)*

4.5.2. *Neural network-based modeling*

To establish a useful relationship between tool wear and cutting parameters a multilayer perceptron (MLP) type neural network was selected. The neural network has two layers: a hidden layer and an output layer. The hidden layer uses a transference function of sigmoid type:

$$f(\mathbf{x}) = \frac{1}{1+\exp[-(b+\sum w_i x_i)]} \qquad [4.11]$$

On the contrary, the output layer uses a linear function:

$$f(\mathbf{x}) = b + \sum w_i x_i \qquad [4.12]$$

To carry out the training process three points, homogenously located, were selected from the experimental data to form the validation set. These are points 6, 16 and 20. The remaining data form the training set. Not only input variables but also output variables were normalized in the range [0, 1] in order to facilitate the neural networks training process.

The network was trained by using the gradient descendent with momentum back-propagation algorithm. In this algorithm, four parameters must be tuned: learning rate *LR*, moment constant *MC*, training epochs *E* and number of hidden nodes *HN*. For this tuning the Taguchi method was used to find the most convenient values for achieving not only lower root mean square error, but also good generalization capabilities, giving the following values: $LR = 0.03$, $MC = 0.7$, $E = 25,000$ and $HN = 3$. It must be remarked that this number of hidden neurons guarantee that there are more training samples (24) than the total amount of free parameters (16), thus the training process is mathematically determined.

Using the aforementioned values, the final training process was carried out. In Figure 4.8, the behavior of the RMSE throughout the training process is shown.

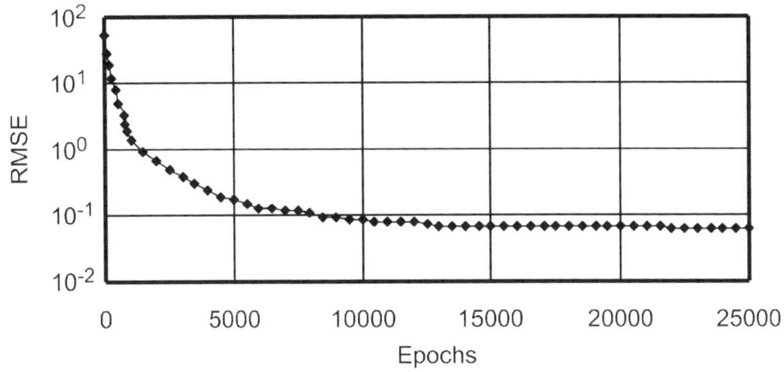

Figure 4.8. *Behavior of the RMSE along the training process [QUI 08] (© 2007, Springer Science and Business Media)*

The trained network achieves a RMSE value of 0.022 and a GC (probability value) equal to 0.670. The obtained model can be implemented by the following C function code:

```
double Wear(double Speed, double Feed, double Time)
{
  double W1[3][3], W2[3], B1[3], B2;
  double X[3], S, L[3];
  int i, j;
  X[0] = (Speed - 80)/(150 - 80);
  X[1] = (Feed - 0.05)/(0.15 - 0.05);
  X[2] = (Time - 5)/(15 - 5);
  W1[0][0]= 1.8753; W1[0][1]= 2.6878; W1[0][2]= 0.9498;
  W1[1][0]=-1.1571; W1[1][1]=-2.8293; W1[1][2]=-1.3605;
  W1[2][0]=-1.6989; W1[2][1]=-1.2206; W1[2][2]=-0.4847;
  W2[0] = -5.4640; W2[1] = -3.5099; W2[2] = -5.7351;
  B1[0] = -3.5050; B1[1] = 3.1480; B1[2] = 3.4430;
  B2 = 9.1174;
  for (i = 0; i <= 2; i++) {
    S = B1[i];
    for (j = 0; j <= 2; j++) {
      S = S + W1[i][j]*X[j];
    }
    L[i] = 1 / (1 + exp(-S));
  }
   S = B2;
  for (i = 0; i <= 2; i++) {
    S = S + W2[i]*L[i];
  }
  return(0.032 + S*(0.333 - 0.032));
}
```

In Figure 4.9 the distribution of residuals for training and validation sets are shown. Analyzing the sets of residuals, it can be remarked that the confidence interval for the difference between the means, which extends from -0.016 to 0.0071, contains the value 0. Therefore, there is not a statistically significant difference between the means of the two samples. Furthermore, the confidence interval for the ratio of the variances of both sets extends from 0.00685372 to 1.17607. Since the interval contains the value 1, there is not a statistically significant difference between the standard deviations of the two samples. Finally, by performing a Kolmogorov-Smirnov test, no statistically significant difference between the two distributions was found. All the above-mentioned tests were carried out at the 95% confidence level.

In Figure 4.10 the neural network model is graphically shown. As can be noted, the surfaces corresponding to different levels of cutting speed have a complex form, which corresponds to the experimental data.

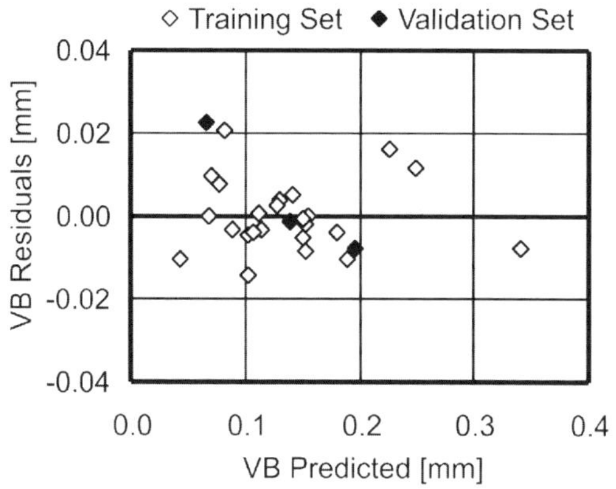

Figure 4.9. *Residuals vs. predicted values for training and validation sets in the neural network model [QUI 08] (© 2007, Springer Science and Business Media)*

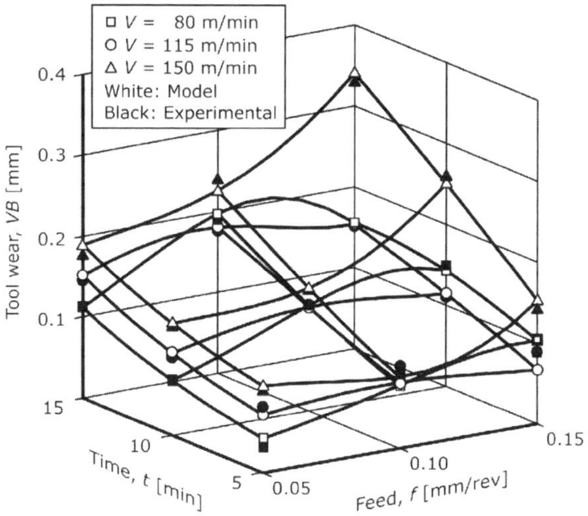

Figure 4.10. *Graphical representation of neural network model [QUI 08] (© 2007, Springer Science and Business Media)*

4.5.3. *Comparison between statistical and neural network models*

In order to evaluate the convenience of using a neural network based model, its outputs are compared with a statistical regression. Different regressions (linear, quadratic and potential) were tried in order to select the most correct model. The obtained equations were:

$$VB = -0.1309 + 9.873 \cdot 10^{-4} V + 0.5878 f + 9.711 \cdot 10^{-2} \tau \qquad [4.13a]$$

$$VB = 0.2939 + 2.168 \cdot 10^{-5} V^2 - 4.656 \cdot 10^{-3} V + 6.573 \cdot 10^{-4} V f \tau \qquad [4.13b]$$

$$VB = 2.020 \cdot 10^{-3} \cdot V^{0,7737} \cdot f^{0.4154} \cdot \tau^{0.6700} \qquad [4.13c]$$

The respective R-squared values were 0.74, 0.79 and 0.79, therefore quadratic and potential models have a higher R-squared statistic than the linear one, however, it is less than 0.8. This indicates that the adjusted models explain less than 80% of the variability in the dependent variable (VB).

In order to select the best statistical model, residual normality was analyzed using four tests: chi-square, Shapiro-Wilks, standardized skewness and standardized kurtosis. For each one, the respective probability value (p-value) was calculated. For the quadratic model the lowest p-value amongst the performed tests was equal to 0.3721. Consequently, the hypothesis that residuals come from a normal distribution cannot be rejected with 90% or higher confidence. On the contrary, for the potential model the lowest p-value is 0.0800, so, the hypothesis that residuals come from a normal distribution can be rejected with 90% confidence. Since residuals from quadratic models seem to have a normal distribution, and residuals from potential models seem to have not, the quadratic model was selected as the most convenient to describe the wear.

Nevertheless, as can be seen in Figure 4.11, where this model is graphically shown, there is a serious divergence between experimental data and predicted values, for several points.

For the neural network model, an R-square statistic equal to 0.98 was obtained. This value is significantly greater than that obtained for the statistical model (0.79). The mean absolute error was 0.0070 for the neural network model while it was 0.0231 for the statistical one. The maximum absolute errors were 0.0633 and 0.0228, respectively.

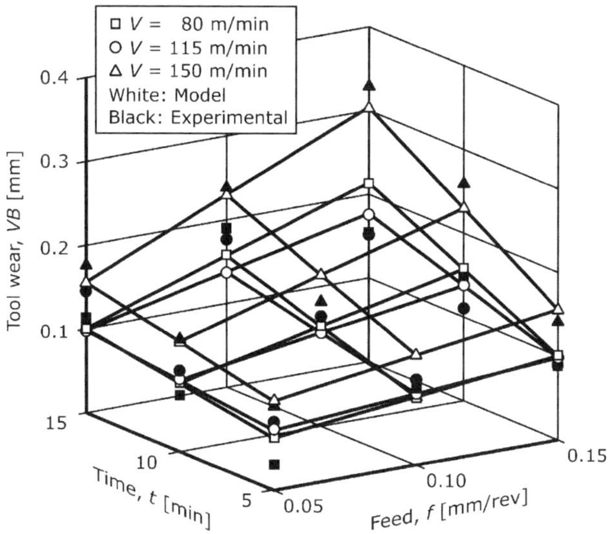

Figure 4.11. *Graphical representation of quadratic statistical model [QUI 08] (© 2007, Springer Science and Business Media)*

In Figure 4.12, residuals are plotted versus predictions, for both models. As can be seen, the neural network model has a less dispersed distribution of residuals.

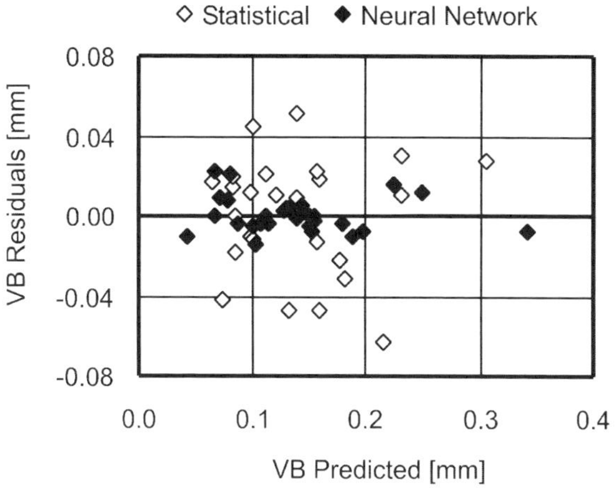

Figure 4.12. *Residuals vs. predicted values for statistical and neural network models [QUI 08] (© 2007, Springer Science and Business Media)*

Furthermore, comparing the graphical representation of both models (see Figures 4.10 and 4.11), it can be noted that surfaces from neural network models have a more closed shape than those expected from the experimental data (Figure 4.7).

Considering everything that has been discussed, we conclude that the neural network model describes the relationship between the tool wear and the cutting speed, feed and cutting time with more precision.

4.6. Concluding remarks

As shown, artificial intelligence techniques have been widely applied in machining process modeling. The most commonly used techniques are artificial neural networks, particularly, the multilayer perceptron. The popularity of this paradigm is due to its relative simplicity, the existence of abundant references and the availability of several user-friendly and well documented implementations. However, other tools such as fuzzy logic, neuro-fuzzy systems and hidden Markov models have gained space in this field.

Artificial intelligence techniques have proved advantageous in machining systems and process modeling, dealing with complex, non-linear problems. These methods have not only shown their effectiveness in pattern recognition, but also in functional approximation for developing on-line and off-line models.

In spite of the above-mentioned profits, the intelligent techniques have important drawbacks mainly caused by bad implementation and poor understanding of their principles and theoretical fundamentals. For example, in neural networks implementation often uses an inadequate quantity of data for training. If the amount of connections between neurons is greater than the available data points, training process, although still possible, is mathematically undetermined. This is a very common problem in neural network-based machining modeling.

Another shortcoming often found in the reviewed papers is the absence of an independent data set to test the generalization capability of the trained models, which is a generally accepted requirement to validate the soft computing models. Moreover, statistical analyses of intelligent model outputs are frequently carried out in a very limited way or not at all.

Finally, it must be mentioned that only a few works report comparisons between proposed artificial intelligent techniques and other more simple tools, such as regressions and polynomials interpolations. Therefore, the pertinence and convenience of applying artificial intelligence techniques are not fully proved.

Nevertheless, if carefully applied, artificial intelligence is a powerful and versatile tool and can provide accurate results in machining process modeling. An open door for future research in this branch is the combination of analytical principles and intelligent techniques to obtain more effective models.

4.7. Acknowledgment

The authors would like to acknowledge Springer Science and Business Media for its kind permission to revise text, tables and figures from previously published material [QUI 08].

4.8. References

[ALA 06] ALANIZ P.D., GÓMEZ R.A., ROMERO R.J., PENICHE R.R., JUREGUI J.C., HERRERA G., "Sensorless detection of tool breaking in milling", *Machining Science and Technology*, vol. 10, no. 2, 2006, p. 263-274.

[BAG 06] BAGCI E., ISIK B., "Investigation of surface roughness in turning unidirectional GFRP composites by using RS methodology and ANN", *International Journal of Advanced Manufacturing Technology*, vol. 31, no. 1-2, 2006, p. 10-17.

[BAS 07] BASAK S., DIXIT U.S., DAVIM J.P., "Application of radial basis function neural networks in optimization of hard turning of AISI D2 cold-worked tool steel with a ceramic tool", *Journal of Engineering Manufacture*, vol. 221, no. 6, 2007, p. 987-998.

[CHE 07] CHEN X., LI B., "Acoustic emission method for tool condition monitoring based on wavelet analysis", *International Journal of Advanced Manufacturing Technology*, vol. 33, no. 9-10, 2007, p. 968-976.

[CHI 00] CHILDS T., MEKAWA K., OBIKAWA T., YAMANO Y., *Metal Machining Theory and Applications*, New York (USA), John Wiley & Sons, 2000, ISBN 0-470-39245-2.

[CHI 07] CHING-KAO C., LU H.S., "The optimal cutting-parameter selection of heavy cutting process in side milling for SUS304 stainless steel", *International Journal of Advanced Manufacturing Technology*, vol. 34, no. 5-6, 2007, p. 440-447.

[CHO 05] CHO S., ASFOUR S., ONAR A., KAUNDINYA N., "Tool breakage detection using support vector machine learning in a milling process", *International Journal of Machine Tools and Manufacture*, vol. 45, no. 3, 2005, p. 241-249.

[CUS 06] CUS F., ZUPERL U., MILFELNER M., "Dynamic neural network approach for tool cutting force modelling of end milling operations", *International Journal of General Systems*, vol. 35, no. 5, 2006, p. 603-618.

[DIN 03] DINI G., "On-line prediction of delamination in drilling of GFRP by using a neural network approach", *Machining Science and Technology*, vol. 7, no. 3, 2003, p. 295-314.

[DON 06] DONG J., SUBRAHMANYAM K.V.R., WONG Y.S., HONG G.S., MOHANTY A.R., "Bayesian-inference-based neural networks for tool wear estimation", *International Journal of Advanced Manufacturing Technology*, vol. 30, no. 9-10, 2006, p. 797-807.

[ENE 03] ENEMOUH E.U., EL GIZAWY A.S., "Optimal neural network model for characterization of process-induced damage in drilling carbon fiber reinforced epoxy composites", *Machining Science and Technology*, vol. 7, no. 3, 2003, p. 389-400.

[FU 06] FU P., HOPE A.D., "The application of B-spline neurofuzzy networks for condition monitoring of metal cutting tool", *PRICAI 2006: Trends in Artificial Intelligence, 9th Pacific Rim International Conference on Artificial Intelligence*, Guilin, China, August 7-11, 2006, p. 1078-1082.

[GIO 03] GIORLEO G., TETI R., PRISCO U., D'ADDONA D., "Merging neural network material rheological behaviour modelling with FEM simulation of orthogonal metal cutting", *Machining Science and Technology*, vol. 7, no. 3, 2003, p. 401-417.

[HSU 08] HSUEH Y.-H., YANG C.-Y., "Prediction of tool breakage in face milling using support vector machine", *International Journal of Advanced Manufacturing Technology*, vol. 37, no. 9-10, 2008, p. 872-880.

[JEM 06] JEMIELNIAK K., BOMBINSKI S., "Hierarchical strategies in tool wear monitoring", *Journal of Engineering Manufacture*, vol. 220, no. 3, 2006, p. 375-381.

[KAN 06] KANG Y., CHANG C.-W., CHU M.-H., CHANG Y.-P., WANG Y.-P., "Estimation of thermal deformation in machine tools using the hybrid autoregressive moving-average - neural network model", *Proceedings of the Institution of Mechanical Engineers, Part B: Journal of Engineering Manufacture*, vol. 220, no. 8, 2006, p. 1317-1323.

[KAS 06] KASSIM A.A., MIAN Z., MANNAN M.A., "Tool condition classification using hidden Markov model based on fractal analysis of machined surface textures", *Machine Vision and Applications*, vol. 17, 2006, p. 327-336.

[KIR 06] KIRBY E.D., CHEN J.C., ZHANG J.Z., "Development of a fuzzy-nets-based in-process surface roughness adaptive control system in turning operations", Expert Systems with Applications, vol. 30 no. 4, 2006, p. 592-604.

[KON 07] KONG F., YU J., "Study of fuzzy stochastic limited cutting width on chatter", *International Journal of Advanced Manufacturing Technology*, vol. 33, no. 7-8, 2007, p. 677-683.

[LEE 07] LEE K.-J., LEE T.-M., YANG M.-Y., "Tool wear monitoring system for CNC end milling using a hybrid approach to cutting force regulation", *International Journal of Advanced Manufacturing Technology*, vol. 32, no. 1-2, 2007, p. 8-17.

[LI 00] LI X., DONG S., VENUVINOD P.K., "Hybrid learning for tool wear monitoring", *International Journal of Advanced Manufacturing Technology*, vol. 16, no. 5, 2000, p. 303-307.

[LI 02] LI X., "A brief review: acoustic emission method for tool wear monitoring during turning", *International Journal of Machine Tools and Manufacture*, vol. 42, no. 2, 2002, p. 157-165.

[LIA 04a] LIANG M., YEAP T., HERMANSYAH A., "A fuzzy system for chatter suppression in end milling", *Journal of Engineering Manufacture*, vol. 218, no. 4, 2004, p. 403-417.

[LIA 04b] LIANG S.Y., HECKER R.L., LANDERS R.G., "Machining process monitoring and control: the state-of-the-art", *Journal of Manufacturing Science and Engineering*, vol. 126, no. 2, 2004, p. 297-310.

[LIA 06a] LIAO T.W., HUA G., QU J., BLAU P.J., "Grinding wheel condition monitoring with hidden Markov model-based clustering methods", *Machining Science and Technology*, vol. 10, no. 4, 2006, p. 511-538.

[MAC 03] MACKERLE J., "Finite element analysis and simulation of machining: an addendum. A bibliography", *International Journal of Machine Tools and Manufacture*, vol. 43, no. 1, 2003, p. 103-114.

[MUK 06] MUKHERJEE I., RAY P.K., "A review of optimization techniques in metal cutting processes", *Computers and Industrial Engineering*, vol. 50, no. 1, 2006, p. 15-34.

[NAT 07] NATARAJAN U., PERIASAMY V.M., SARAVANAN R., "Application of particle swarm optimisation in artificial neural network for the prediction of tool life", *International Journal of Advanced Manufacturing Technology*, vol. 31, no. 9-10, 2007, p. 871-876.

[ONW 08] ONWUBOLU G.C., "Modelling and predicting surface roughness in turning operations using hybrid differential evolution and the group method of data handling networks", *Journal of Engineering Manufacture*, vol. 222, no. 7, 2008, p. 785-795.

[OZE 07a] ÖZEL T., KARPAT Y., "Multi-objective optimization for turning processes using neural network modeling and dynamic-neighborhood particle swarm optimization", *International Journal of Advanced Manufacturing Technology*, vol. 35, no. 3-4, 2007, p. 234-247.

[OZE 07b] ÖZEL T., KARPAT Y., FIGUEIRA L., DAVIM J.P., "Modelling of surface finish and tool flank wear in turning of AISI D2 steel with ceramic wiper inserts", *Journal of Materials Processing Technology*, vol. 189, no. 1-3, 2007, p. 192-198.

[PAN 07] PANDA S.S., CHAKRABORTY D., PA S.K., "Monitoring of drill flank wear using fuzzy back-propagation neural network", *International Journal of Advanced Manufacturing Technology*, vol. 34, no. 3-4, 2007, p. 227-235.

[PAT 07] PATRA K., PAL S.K., BHATTACHARYYA K., "Application of wavelet packet analysis in drill wear monitoring", *Machining Science and Technology*, vol. 11, no. 3, 2007, p. 413-432.

[QUI 08] QUIZA R., FIGUEIRA L., DAVIM J.P., "Comparing statistical models and artificial neural networks on predicting the tool wear in hard machining D2 AISI steel", *International Journal of Advanced Manufacturing Technology*, vol. 37, no. 7-8, 2008, p. 641-648.

[RAO 06] RAO C.H.S., RAO D.N., RAO R.N.S., "Online prediction of diffusion wear on the flank through tool tip temperature in turning using artificial neural networks", *Journal of Engineering Manufacture*, vol. 220, no. 12, 2006, p. 2069-2076.

[REH 05] REHORN A.G., JIANG J., ORBAN P.E., BORDATCHEV E.V., "State-of-the-art methods and results in tool condition monitoring: a review", *International Journal of Advanced Manufacturing Technology*, vol. 26, no. 7-8, 2005, p. 693-710.

[SAN 06] SANJAY C., JYOTHI C., "A study of surface roughness in drilling using mathematical analysis and neural networks", *International Journal of Advanced Manufacturing Technology*, vol. 29, no. 9-10, 2006, p. 846-852.

[SAR 07] SARMA D.K., DIXIT U.S., "A comparison of dry and air-cooled turning of grey cast iron with mixed oxide ceramic tool", *Journal of Materials Processing Technology*, vol. 190, no. 1-3, 2007, p. 160-172.

[SEG 06] SEGONDS S., REDONNET J.-M., BÈS C., LANDON Y., LAGARRIGUE P., "Characterisation of the workpiece dilatation phenomenon during machining using the neural network method. Application to NC turning", *International Journal of Advanced Manufacturing Technology*, vol. 27, no. 1-2, 2006, p. 83-89.

[SHA 07] SHARMA V., SHARMA S., SHARMA A., "An approach for condition monitoring of a turning tool", *Proceedings of the Institution of Mechanical Engineers, Part B: Journal of Engineering Manufacture*, vol. 221, no. 4, 2007, p. 635-648.

[SHE 07] SHEIKH-AHMAD J., TWOMEY J., KALLA D., LODHIA P., "Multiple regression and committee neural network force prediction models in milling FRP", *Machining Science and Technology*, vol. 11, no 3, 2007, p. 391-412.

[SIC 02] SICK B., "On-line and indirect tool wear monitoring in turning with artificial neural networks: a review of more than a decade of research", *Mechanical Systems and Signal Processing*, vol. 16, no. 4, 2002, p. 487-546.

[SIL 06] SILVA R.G., WILCOX S.J., REUBEN R.L., "Development of a system for monitoring tool wear using artificial intelligence techniques", *Journal of Engineering Manufacture*, vol. 220, no. 8, 2006, p. 1333-1346.

[SIN 06] SINGH A.K., PANDA S.S., CHAKRABORTY D., PAL S.K., "Predicting drill wear using an artificial neural network", *International Journal of Advanced Manufacturing Technology*, vol. 28, no. 5-6, 2006, p. 456-462.

[SON 06] SONAR D.K., DIXIT U.S., OJHA D.K., "The application of a radial basis function neural network for predicting the surface roughness in a turning process", *International Journal of Advanced Manufacturing Technology*, vol. 27, no. 7-8, 2006, p. 661-666.

[THA 06] THANGAVEL P., SELLADURAI V., SHANMUGAM R., "Application of response surface methodology for predicting flank wear in turning operation", *Journal of Engineering Manufacture*, vol. 220, no. 6, 2006, p. 997-1003.

[WAN 05] WANG S.-G., HSU Y.-L., "One-pass milling machining parameter optimization to achieve mirror surface roughness", *Journal of Engineering Manufacture*, vol. 219, no. 1, 2005, p. 177-181.

[WAN 06] WANG X., CHEN P., TANSEL I.N., YENILMEZ A., "Transformations in machining. Part 1. enhancement of wavelet transformation neural network (WT-NN) combination with a preprocessor", *International Journal of Machine Tools & Manufacture*, vol. 46, no. 1, 2006, p. 36-42.

[WAN 08] WANG X., WANG W., HUANG Y., NGUYEN N., KRISHNAKUMAR K., "Design of neural network-based estimator for tool wear modeling in hard turning", *Journal of Intelligent Manufacturing*, vol. 19, no. 4, 2008, p. 383-396.

[YU 06] YU G., QUI H., DJURDJANOVIC D., LEE J., "Feature signature prediction of a boring process using neural network modeling with confidence bounds", *International Journal of Advanced Manufacturing Technology*, vol. 30, no. 7-8, 2006, p. 614-621.

[YUM 06] YUMAK O., ERTUNC H.M., "Tool wear condition monitoring in drilling processes using fuzzy logic", *Proceedings of the 13th International Conference*, ICONIP 2006, Hong Kong, China, October 3-6, 2006., p. 508-517.

Chapter 5

Design of Experiments

This chapter deals with various techniques to plan the experimental layout in order to reduce cost and time. Furthermore, the modeling and optimization of machining processes are discussed.

5.1. Introduction

In any machining process, the primary goal is to obtain the required level of quality characteristics of the finished product. This is brought about by proper settings of the control factors and minimizing the effect of uncontrolled factors on the machining processes. This necessitates the identification of quality characteristics of the product and various controllable and noise factors which affect the machining. Furthermore, it is necessary to establish a relationship between the quality characteristics and control factors in order to analyze the machining behavior, which calls for the experimentation on the machining process. This involves the measurement of quality characteristics of the product/process for various combinations of controlled factors which are influenced by noise factors. These measurements will help to understand the machining behavior on which decisions can be drawn in order to properly set the control factors so as to obtain the required quality level and minimize the variance due to noise factors.

Chapter written by Vinayak N. GAITONDE, Ramesh S. KARNIK and J. Paulo DAVIM.

5.1.1. *Conventional approach*

Normally, in machining experiments, each factor is defined with a finite number of settings, known as "*levels*", within the selected range of the operation. In the conventional experimentation approach, each factor is varied at a time over the defined levels, while the other factors are held constant at various combinations of level settings. Suppose there are *k*-number of control factors to be studied in a machining process and each defined with *l* number of levels. In order to completely characterize the process, it requires l^k number of experiments to be performed. However, if it is required to study a large number of factors, each defined with more number of levels, the total number of experiments becomes too huge, which adds to the cost and time of experimentation. This is overcome by planning the experimental layout based on "*design of experiments*" (DOE) [MON 01].

There are several types of DOE available in the literature, which are based on statistical theory. The selection of proper DOE mainly depends on the purpose of experimentation, which includes the development of mathematical modeling and/or the optimization of the machining processes.

5.1.2. *Need for mathematical modeling*

In any machining process, a mathematical model describes the relationship of a quality characteristic with the control factors. Consider the drilling of a carbon fiber reinforced plastic (CFRP) composite using a cemented carbide tool. The drilling of CFRP is significantly affected by the delamination tendency, which leads to the damage at the entrance of drilled hole [GAI 08a]. The delamination significantly reduces the performance and aesthetical aspect of final product. The quality characteristic of drilling CFRP is characterized by "*delamination factor* (F_d)", which is defined as the ratio of maximum diameter of the damaged zone (D_{max}) to the diameter of the drill (D_0), as shown in Figure 5.1.

The investigations revealed that the delamination can be controlled by cutting speed (v), feed rate (f) and point angle of drill (θ). In order to study the behavior of the delamination factor as a function of cutting speed, feed rate and point angle as the control factors, drilling experiments on CFRP laminates need to be conducted. This will enable the engineer to make the decisions on control factor settings in order to improve the quality of the drilled hole. In addition, the investigation also helps to study the interaction effects due to cutting speed, feed rate and point angle on delamination tendency. The second order mathematical model of delamination factor is of the form:

$$F_d = b_0 + b_1 v + b_2 f + b_3 \theta + b_{12} vf + b_{13} v\theta + b_{23} f\theta + b_{11} v^2 + b_{22} f^2 + b_{33} \theta^2 \qquad [5.1]$$

where $b_0, \ldots, b_{33}$ are the coefficients of the polynomials. The mathematical model enables us to predict and analyze the quality characteristic as a function of control factors, optimize the factors for getting the desired quality and control the machining process.

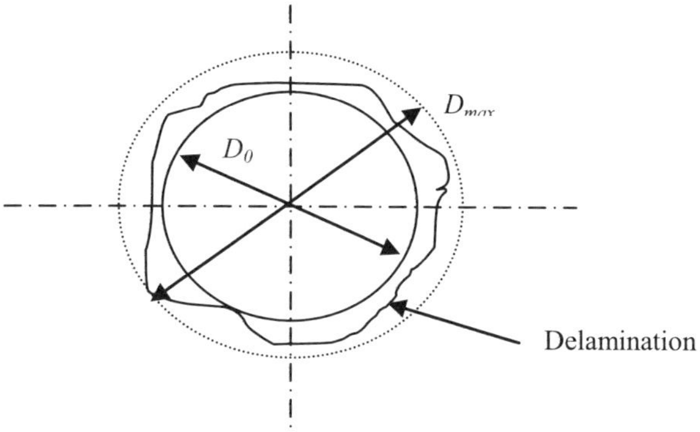

Figure 5.1. *Delamination factor scheme for CFRP composites drilling (adapted from Gaitonde et al. 2008, with permission from Elsevier [GAI 08a])*

5.2. Design of experiments

For the development of the mathematical models based on experimental data, it is essential to carefully plan the experiments. It is possible to plan the experiments using the DOE approach so as to reduce the number of experiments, which in turn reduces the time and cost involved in the experimentation [MON 01]. The experimental design systematically defines the efficient set of experimental sampling points at which the responses must be computed or observed. The DOE takes levels of input variables to formulate the different combinations at which the outputs are observed or computed.

5.2.1. *Classification of DOE*

There are many types of experimental designs, which are mainly based on the number of factors and their levels. The appropriate DOE to be selected is based on the following criteria:

– purpose: model development/optimization;

– nature of model: linear or non-linear models to be developed;

– number of factors and the number of levels for each factor.

Based on the above criteria, the DOE can be classified as follows [COC 92; MON 01]:

– Full factorial design: in this design, all possible combinations of levels of the factors are considered in each complete trial or replication of the experiments. The number of combinations of the experiments are ($l_a\ l_b ... l_n$), where, $l_a\ l_b\ ... l_n$ are the number of levels for factors a, $...n$. This design can be used to develop linear or non-linear regression models of the quality characteristics of any machining process.

– Fractional factorial design: this design consists of only a fraction of complete factorial experimental design. This design is based on the assumption of negligible higher order interaction. Furthermore, these designs enable us only to determine the main effects and lower order interaction effects. Nevertheless, this design makes it possible to develop linear or non-linear regression models of the machining process.

– Taguchi orthogonal arrays: these are the special kind of fractional factorial designs with reduced number of experiments. These arrays are mainly used for the process optimization and to analyze the main and interaction effects of various factors on the quality characteristics. Orthogonal arrays are extensively used in the robust design which aims at minimizing the effect of noise factors on variation of quality characteristic.

– Central composite design: this design is used to develop the second order models. Normally, central composite design is made up of cube points, star points and center points.

– Box-Behnken design: this is a 3-level spherical design, which can be employed to develop non-linear process models.

5.2.1.1. *DOE based on full factorial design (FFD)*

Suppose that a second order mathematical model is to be developed to predict the delamination factor at the entrance of a drilled hole in CFRP composite [GAI 08a]. Three levels are defined for each of the control factors, cutting speed (v), feed rate (f) and point angle of drill (θ) as summarized in Table 5.1. For three factors, with three levels defined for each of the factors, 27 experiments are performed on CFRP composites as per FFD. The delamination damage is measured and the delamination factor is calculated for each combination of the factor levels and is illustrated in Table 5.2. In order to avoid systematic errors, the experiment trials need to be randomized. The second order model to predict the delamination factor can be developed based on the above experimental investigations. The main drawback of FFD is that the number of experiments increases with the increase in the number of factors or their levels. However, the FFD design allows the development of accurate models. Hence, FFD are mainly limited to cases with a smaller number of factors.

Factor	Symbol	Unit	Level		
			1	2	3
Cutting speed	v	m/min	60	120	600
Feed rate	f	m/min	1	3	6
Point angle	θ	degree	85	115	130

Table 5.1. *Factors and levels of machining parameters in CFRP composite drilling (adapted from Gaitonde et al. 2008, with permission from Elsevier [GAI 08a])*

Trial no.	Design parameters			Delamination factor (F_d)
	v (m/min)	f (m/min)	θ (degree)	
1	60	1	85	1.0380
2	60	3	85	1.5340
3	60	6	85	1.5925
4	120	1	85	1.0365
5	120	3	85	1.0765
6	120	6	85	1.3270
7	600	1	85	1.0335
8	600	3	85	1.0745
9	600	6	85	1.3385
10	60	1	115	1.5810
11	60	3	115	1.9430
12	60	6	115	2.1360
13	120	1	115	1.1395
14	120	3	115	1.7715
15	120	6	115	2.2095
16	600	1	115	1.3390
17	600	3	115	1.3525
18	600	6	115	1.5630
19	60	1	130	1.9070
20	60	3	130	1.6285
21	60	6	130	2.2020
22	120	1	130	1.6550
23	120	3	130	1.7800
24	120	6	130	2.3565
25	600	1	130	1.0865
26	600	3	130	1.4325
27	600	6	130	2.2800

Table 5.2. *Layout plan and delamination factor in CFRP composite drilling (adapted from Gaitonde et al. 2008, with permission from Elsevier [GAI 08a])*

5.2.1.2. *DOE based on central composite design (CCD)*

An unmanageable number of experiments required in the FFD for a large number of factors leads to an introduction of the designs that require fewer design points, while maintaining an acceptable accuracy in the prediction.

Consider a case of second order burr model development in the drilling of AISI 316 stainless steel. The burr is characterized by burr height. The control factors considered are cutting speed (v), feed (f), drill diameter (d), point angle (θ) and lip clearance angle (ψ). A five level, half replicate, second order, rotatable CCD [COC 92] was used to study linear, quadratic and two factor interaction effects between the five factors and the responses. The rotatability in the designs ensures that the variance of the estimated response is constant at a fixed distance from the center point. The upper limit of a factor was coded as +2 and the lower limit as –2, coded values for intermediate values being calculated from the following relationship:

$$X_i = \frac{2[2X-(X_{max}+X_{min})]}{X_{max}-X_{min}} \quad [5.2]$$

where X_i: required coded value of a factor X; X: any value of the factor from X_{min} to X_{max}; X_{min} and X_{max}: lower and upper levels of the factor X. The experimental factors and their levels are summarized in Table 5.3.

Factor	Notation		Unit	Level				
	Natural	Coded		-2	-1	0	1	2
Cutting speed	v	X_1	m/min	8	12	16	20	24
Feed	f	X_2	mm/rev	0.04	0.06	0.08	0.10	0.12
Drill diameter	d	X_3	mm	4	10	16	22	28
Point angle	θ	X_4	degree	118	122	126	130	134
Lip clearance angle	ψ	X_5	degree	8	9	10	11	12

Table 5.3. *Experimental factors and levels selected in drilling of AISI 316 stainless steel material*

The design comprises a half replication of 2^{5-1} (=16) fractional factorial design (cube points), 6-center points and 10-star points [COC 92]. The matrix of second order CCD consists of 32 sets of coded conditions as illustrated in Table 5.4. All the factors at the intermediate level (0) constitute the center points and the combinations of each of the drilling factors either at its lowest (-2) or highest (+2) with the other four factors of the intermediate levels constitute the star points.

Run no.	X_1	X_2	X_3	X_4	X_5	Burr height (mm)
1	-1	-1	-1	-1	+1	1.189
2	+1	-1	-1	-1	-1	0.302
3	-1	+1	-1	-1	-1	0.628
4	+1	+1	-1	-1	+1	0.723
5	-1	-1	+1	-1	-1	0.465
6	+1	-1	+1	-1	+1	0.141
7	-1	+1	+1	-1	+1	0.018
8	+1	+1	+1	-1	-1	0.313
9	-1	-1	-1	+1	-1	0.533
10	+1	-1	-1	+1	+1	1.386
11	-1	+1	-1	+1	+1	0.195
12	+1	+1	-1	+1	-1	0.411
13	-1	-1	+1	+1	+1	0.479
14	+1	-1	+1	+1	-1	0.539
15	-1	+1	+1	+1	-1	0.698
16	+1	+1	+1	+1	+1	0.816
17	-2	0	0	0	0	0.562
18	+2	0	0	0	0	0.726
19	0	-2	0	0	0	1.693
20	0	+2	0	0	0	1.164
21	0	0	-2	0	0	0.837
22	0	0	+2	0	0	0.309
23	0	0	0	-2	0	0.402
24	0	0	0	+2	0	0.788
25	0	0	0	0	-2	0.261
26	0	0	0	0	+2	0.457
27	0	0	0	0	0	1.184
28	0	0	0	0	0	1.239
29	0	0	0	0	0	1.345
30	0	0	0	0	0	1.367
31	0	0	0	0	0	1.234
32	0	0	0	0	0	1.357

Table 5.4. *Design matrix of CCD and measured values of burr height in drilling of AISI 316 stainless material*

The trials indicate the sequence number of the run under consideration. X_1, X_2, X_3, X_4 and X_5 represent the notation used for controllable factors as defined in Table 5.3. The measured values of burr height are summarized in Table 5.4. In Table 5.4, the run numbers from 1 to 16 correspond to cube points, 17 to 26 are star points and 27 to 32 represent center points. The replications of center points provide a means for estimating pure experimental error. In this case, the experimental layout with CCD is advantageous since it is possible to develop a non-linear mathematical model with only 32 experiments. On the other hand, for five factors and three levels, it would require 3^5 (= 243) experiments as per FFD.

5.3. Modeling using response surface methodology

The response surface methodology (RSM) is a collection of statistical and mathematical techniques useful for developing, improving and optimizing the processes that provide an overall perspective of the system response within the design space [MON 01]. The predictive capability of RSM is greatly influenced by the distribution of sampling points in the design space. In order to select the design points for training, the DOE techniques are applied. The RSM refers not simply to the use of a response surface as a multivariate function but also to the process for determining the polynomial coefficients themselves. A response surface equation is a polynomial regression to a data set. The process is straightforward if a sufficiently large data set is available, i.e., if the number of members in the data set is at least as large as the number of coefficients in the polynomial. The DOE technique provides the required basis for this critical step in the methodology.

5.3.1. *Steps for the development of an RSM-based model*

The construction of RSM-based models generally involves three steps: (i) choosing the polynomial equation for the representation of the model, (ii) model fitting to the measured/simulated data and (iii) statistical validation for the response surface.

5.3.1.1. *Choosing the polynomial equation for the representation of a model*

After performing a set of experiments to obtain the process responses or outputs according to the experimental designs, the next step is to take the vectors of inputs *(X)* and the corresponding outputs (*Y*) to fit the appropriate model. The most widely used response surface function is a mathematical polynomial function. A typical response surface model limits the order of polynomial to one or two since the low-degree models contain fewer terms than the higher-degree models and thus require fewer experiments to be performed. If the responses exhibit non-linear behavior, the second-order polynomial is considered as a response surface model. The response function considering a second-order polynomial model is as follows [COC 92; MON 01]:

$$Y_u = b_0 + \sum_{i=1}^{k} b_i X_{iu} + \sum_{i=1}^{k-1} \sum_{j=i+1}^{k} b_{ij} X_{iu} X_{ju} + \sum_{i=1}^{k} b_{ii} X_{iu}^2 \quad [5.3]$$

where Y_u: response under study, X_{iu}: coded values of the ith variables for u^{th} experiments, b_o, b_i, b_{ij}, b_{ii}: regression coefficients to be determined; k: number of input factors.

In equation [5.3], the first term b_0 represents the response at the center of the experiment when all the factors are at 0 levels. The second term under the summation sign of a polynomial equation is attributable to linear effect, whereas the third term corresponds to interactive effects. The fourth term of the equation includes the higher-order effect of factors. In the RSM, each X-factor can be coded so that the upper level of X is taken as $+\alpha$ and the lower level as $-\alpha$ in order to design the experiments in an optimized way.

5.3.1.2. *Model fitting to the measured/simulated data*

The values of regression coefficients of linear, quadratic and interaction terms of the model are determined by [MON 01]:

$$b = (X^T X)^{-1} X^T Y \quad [5.4]$$

where b: regression coefficients, X: calculation matrix which contains the coded values of main and interaction terms, X^T: transpose of X: $(X^T X)^{-1}$ inverse matrix of a $(X^T X)$ and Y: matrix of measured/simulated response.

5.3.1.3. *Statistical validation for the response surface*

The analysis of variance (ANOVA) [MON 01] technique is used to check the adequacy of the developed model for the desired confidence interval. The ANOVA table includes the sum of squares (*SS*), degrees of freedom (*DF*) and mean square (*MS*). In the ANOVA, the contribution for *SS* is from the first order terms, the second order terms and the residual error. The *MS* are obtained by dividing the *SS* of each of the sources of variation by the respective *DF*. The formulae for *SS* and *DF* are as follows:

Source of variation	SS	DF
First order terms:	$SS_f = \sum_{i=1}^{k} b_i(iy)$	k
Second order terms:	$SS_s = b_0(0y) + \sum_{i=1}^{k} b_{ii}(iiy) + \sum_{i<j} b_{ij}(ijy) - \frac{G^2}{N}$	$\frac{k(k+1)}{2}$
Regression:	$SS_r = SS_f + SS_s$	$\frac{k(k+3)}{2}$
Total:	$SS_T = \sum_{u=1}^{N} y_u^2 - \frac{G^2}{N}$	$(N - 1)$
Residual error:	$SS_e = SS_T - SS_r$	$(N-1) - \frac{k(k+3)}{2}$

where G: grand total; N: number of data points; k: number of input factors;

(iy): sum of products of ith column in X with $Y = \sum_{u=1}^{N} x_{iu} y_u$;

(ii): sum of squares of ith column in $X = \sum_{u=1}^{N} x_{iu}{}^2$;

(ij): sum of products of ith and ith columns in $X = \sum_{u=1}^{N} x_{iu} x_{ju}$;

Fisher's ratio of regression is given as:

$$F_r = \frac{(MS)_r}{(MS)_e} \quad [5.5]$$

where $(MS)_r$ is the mean square of regression; $(MS)_e$ is mean square of residual error. As per the ANOVA technique, the model developed is adequate within the specified confidence interval, if F_r exceeds the standard tabulated value of *F-ratio*.

5.3.2. *Response surface modeling – a case study using FFD*

The study presented here describes the development of an RSM-based model to predict the delamination factor at the entrance of a drilled hole in CFRP composites [GAI 08a]. The three control factors and their three levels are given in Table 5.1. The corresponding experimental layout plan as per FFD along with computed values of the delamination factor is presented in Table 5.2. Following the model development procedure as explained in section 5.3.1, the second order RSM-based model of delamination factor in natural form is given as:

$$\begin{aligned} F_d = & -0.810444 - 0.001889v - 0.109957f + 0.034546\,\theta + 0.000011vf \\ & - 0.000009\,v\theta + 0.001670\,f\theta + 0.000003v^2 + 0.005530\,f^2 - 0.000115\,\theta^2 \end{aligned} \qquad [5.6]$$

where v is in m/min; f in m/min; θ in degrees.

The model adequacy is checked through ANOVA and is presented in Table 5.5. Here, the F-ratio of regression is 10.78, which is greater than F-table = 2.49, indicating the adequacy of the developed model at 95% confident interval.

Source of variation	Degrees of freedom	Sum of squares	Mean square	F-ratio
Regression	9	3.91201	0.43467	10.78
Residual error	17	0.68527	0.04031	
Total	26	4.59728		

F-table $_{(9, 17, 0.05)} = 2.49$

Table 5.5. *Summary of ANOVA for RSM based delamination factor model in drilling of CFRP composite (adapted from Gaitonde et al. 2008, with permission from Elsevier [GAI 08a])*

The mathematical model of equation [5.6] can be used to analyze the delamination tendency according to control factors, known as response surface analysis. This comprises responses surfaces, which illustrate the two-factor interaction effect on delamination. The response surface plots of two-factor interaction effects are generated using MINITAB software [MIN 03] and are depicted in Figures 5.2, 5.3 and 5.4.

5.3.3. *Response surface modeling – a case study using CCD*

Consider the development of a burr height model using the experimental layout plan as per CCD. The control factors and their levels are already explained in section 5.2.1.2. The measured burr height as per CCD (Table 5.4) is used to develop a second order RSM-based burr height model. The response surface equation for five factors considered in the present case is of the form:

$$Y = b_0 + \sum_{i=1}^{5} b_i X_i + \sum_{i=1}^{4} \sum_{j=i+1}^{5} b_{ij} X_i X_j + \sum_{i=1}^{5} b_{ii} X_i^2 \qquad [5.7]$$

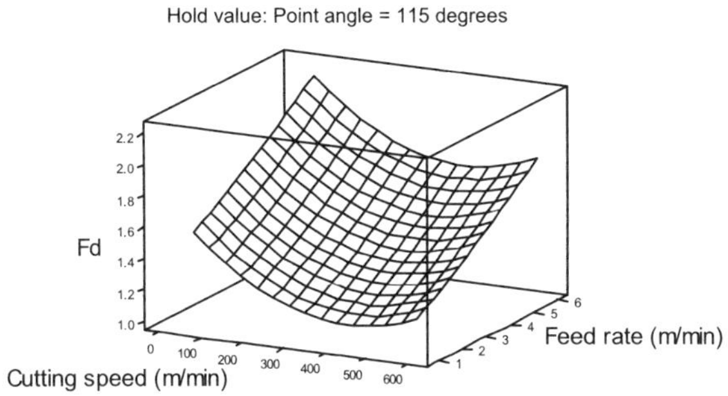

Figure 5.2. *Interaction effect due to cutting speed and feed rate on delamination factor for a point angle of 115^0 (adapted from Gaitonde et al. 2008, with permission from Elsevier [GAI 08a])*

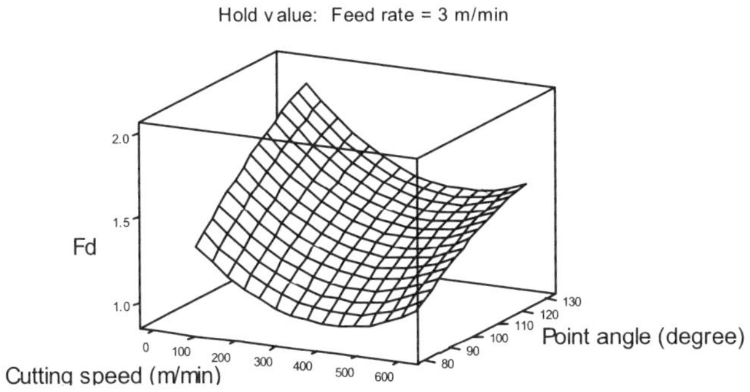

Figure 5.3. *Interaction effect due to cutting speed and point angle on delamination factor for a feed rate of 3 m/min (adapted from Gaitonde et al. 2008, with permission from Elsevier [GAI 08a])*

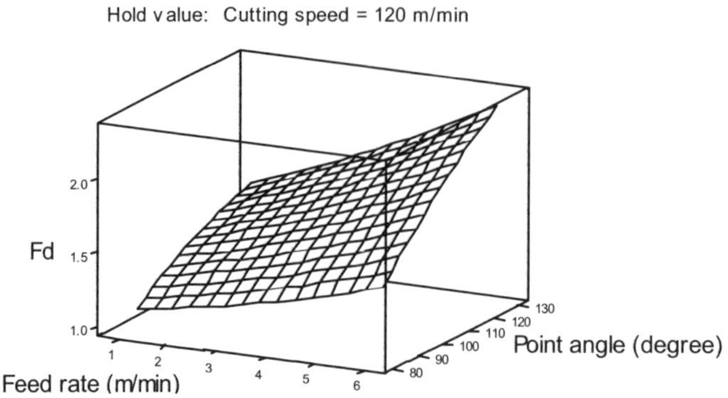

Figure 5.4. *Interaction effect due to feed rate and point angle on delamination factor for a cutting speed of 120 m/min (adapted from Gaitonde et al. 2008, with permission from Elsevier [GAI 08a])*

The mathematical model in coded form as determined by the regression analysis to predict the burr height for drilling AISI 316 stainless steel work pieces is developed as explained in sections 5.3.1 and is given by:

$$\begin{aligned} B_h = {} & 1.29234 + 0.03142X_1 - 0.09542X_2 - 0.12308X_3 + 0.08542X_4 + 0.06042X_5 \\ & + 0.06387X_1X_2 - 0.008X_1X_3 + 0.12925X_1X_4 + 0.1215X_1X_5 + 0.10463X_2X_3 \\ & - 0.02513X_2X_4 - 0.10337X_2X_5 + 0.1195X_3X_4 - 0.13625X_3X_5 + 0.02075X_4X_5 \\ & - 0.16559X_1^2 + 0.03053X_2^2 - 0.18334X_3^2 - 0.17784X_4^2 - 0.23684X_5^2 \end{aligned} \quad [5.8]$$

Using equation [5.2], the coded form of equation [5.8] is converted into natural form, which is given as:

$$\begin{aligned} B_h = {} & -176.191 - 1.041v + 47.548f - 0.322d + 2.587\theta + 4.434\psi + 0.798vf - 0.0003vd \\ & + 0.008v\theta + 0.030v\psi + 0.872fd - 0.314f\theta - 5.1698f\psi + 0.005d\theta - 0.023d\psi \\ & + 0.005\theta\psi - 0.010v^2 + 76.346f^2 - 0.005d^2 - 0.011\theta^2 - 0.237\psi^2 \end{aligned} \quad [5.9]$$

The model adequacy is checked through ANOVA and is presented in Table 5.6. Here, the *F*-ratio of regression is 60.82, which is greater than the *F*-table = 4.95, indicating the adequacy of the developed model at 95% confident interval.

Source of variation	Degrees of freedom	Sum of squares	Mean square	F-ratio
Regression	20	6.04209	0.302105	60.82
Residual error	11	0.05464	0.004967	
Total	31	6.09673		

F-table $_{(6, 5, 0.05)} = 4.95$

Table 5.6. *Summary of ANOVA for RSM based burr height model in drilling of AISI 316 stainless steel*

5.3.4. *Response surface modeling – a case study using BBD*

The principal advantage of Box-Behnken design (BBD) is that it requires fewer design points and does not have axial points as compared to CCD [MON 01]. Considering the modeling of the burr height problem in drilling of AISI 316L stainless steel using an HSS twist drill [GAI 08b], the factors and levels are summarized in Table 5.7. The higher level of a factor was coded as +1, medium level as 0 and the lower level as -1 as illustrated in Table 5.7. The 3 levels, 5-factor BBD experimental plan and the measured burr height are summarized in Table 5.8.

The RSM-based model of burr height can be developed using the procedure as explained in section 5.3.1. This equation in coded form is given by [GAI 08b]:

$$\begin{aligned} B_h = {} & 0.160667+0.006125X_1+0.064125X_2-0.0140625X_3-0.000625X_4+0.0578125X_5 \\ & 0.14325X_1X_2+0.03025X_1X_3-0.01575X_1X_4+0.02825X_1X_5-0.03775X_2X_3 \\ & +0.0485X_2X_4-0.087X_2X_5-0.15825X_3X_4-0.025X_3X_5-0.1145X_4X_5 \\ & -0.01925X_1^2+0.13875X_2^2+0.0495X_3^2+0.0629167X_4^2+0.0911667X_5^2 \end{aligned} \qquad [5.10]$$

Factor	Notation		Unit	Level		
	Natural	Coded		-1	0	1
Cutting speed	v	X_1	m/min	8	16	24
Feed	f	X_2	mm/rev	0.04	0.08	0.12
Drill diameter	d	X_3	mm	10	16	22
Point angle	θ	X_4	degree	118	126	134
Lip clearance angle	ψ	X_5	degree	8	10	12

Table 5.7. *Experimental factors and their levels selected for BBD in drilling of AISI 316L stainless material (adapted from Gaitonde et al. 2008, with permission from Springer [GAI 08b])*

Trial No.	X_1	X_2	X_3	X_4	X_5	Burr height (mm)
1	-1	0	0	-1	0	0.184
2	0	-1	-1	0	0	0.262
3	0	0	-1	-1	0	0.122
4	-1	0	-1	0	0	0.228
5	0	0	0	0	0	0.156
6	0	0	0	1	-1	0.388
7	0	0	0	-1	1	0.478
8	0	-1	0	0	1	0.466
9	1	0	0	0	1	0.333
10	0	0	1	0	-1	0.243
11	0	0	0	0	0	0.161
12	-1	0	0	0	-1	0.189
13	0	0	0	0	0	0.156
14	0	0	-1	1	0	0.444
15	1	0	0	1	0	0.180
16	-1	0	1	0	0	0.146
17	0	-1	0	1	0	0.247
18	0	0	1	1	0	0.111
19	0	1	0	0	-1	0.489
20	0	0	0	0	0	0.161
21	0	0	-1	0	-1	0.229
22	1	0	-1	0	0	0.185
23	-1	1	0	0	0	0.486
24	0	1	1	0	0	0.355
25	0	1	0	-1	0	0.383
26	0	0	0	0	0	0.166
27	0	0	0	1	1	0.264
28	1	0	0	0	-1	0.152
29	-1	0	0	1	0	0.201
30	0	-1	0	0	-1	0.182
31	0	1	-1	0	0	0.469
32	0	0	-1	0	1	0.402
33	-1	0	0	0	1	0.257
34	0	0	1	-1	0	0.422
35	1	0	0	-1	0	0.226
36	0	1	0	1	0	0.469
37	1	-1	0	0	0	0.364
38	0	0	0	0	0	0.164
39	0	0	0	-1	-1	0.144
40	0	0	1	0	1	0.316
41	0	-1	0	-1	0	0.355
42	-1	-1	0	0	0	0.076
43	1	1	0	0	0	0.201
44	0	1	0	0	1	0.425
45	0	-1	1	0	0	0.299
46	1	0	1	0	0	0.224

Table 5.8. *BBD matrix and observed values of burr height in drilling of AISI 316L stainless steel material (adapted from Gaitonde et al. 2008, with permission from Springer [GAI 08b])*

5.4. Taguchi-based designs

The Taguchi robust design principle is based on matrix experiments. The classical experimental design methods are too complex and are not easy to use. If there are more parameters, a large number of experiments have to be carried out. This problem is overcome in the Taguchi method, which uses a special design of orthogonal arrays (OA) [PHA 89] to study the entire parameter space with only a small number of experiments. The OA is a major tool used in the robust design, which is used to study many design parameters by means of a response. The purpose of conducting an experiment based on OA is to determine the optimum level for each factor and to establish the relative significance of the individual factors in terms of their main effects on the response. The OA gives acceptable estimates of factor effects with fewer experiments when compared to the traditional methods.

5.4.1. *Experimental plans and orthogonal arrays*

Depending on the number of factors and its setting levels, a suitable array is selected. Each column of OA designates a factor and its setting levels in each experiment and each row designates an experiment trial with a combination of levels of different factors in that trail. The first step in constructing an OA to fit a specific case study is to count the total degrees of freedom, which gives the minimum number of experiments to be performed. The selection of OA [ROS 96] begins with the distinct number of levels (l) defined for the number of factors (k). The minimum number of trials in the OA is

$$N_{min} = (l-1)k+1 \qquad [5.11]$$

Taguchi has tabulated 18 standard OAs, each comprising N_0 ($\geq N_{min}$) number of trials for different numbers of factors and their levels, which can be directly used for the experimental plan [ROS 96].

Consider the factors and levels given in Table 5.9, which corresponds to drilling experiments on a SUPERPAN DÉCOR (melamine coating layer) medium density fiberboard (MDF) panel using cemented carbide twist drills [GAI 08c]. The standard L_9 OA is used to conduct experiments. Table 5.10 illustrates the 9 trials of various level combinations of feed rate and cutting speed of L_9 OA and the computed values of the delamination factor at the entrance ($F_{d(entry)}$) and exit ($F_{d(exit)}$) of drilled holes.

Code	Factor	Level		
		1	2	3
A	Feed rate (f), m/min	0.10	2.55	5.00
B	Cutting speed (v), m/min	16	63	110

Table 5.9. *Factors and levels selected in drilling of SUPERPAN DÉCOR (melamine coating layer) MDF panel (adapted from Gaitonde et al. 2008, with permission from Elsevier [GAI 08c])*

Trial no.	Levels of input factors		Response	
	A	B	$F_{d(entry)}$	$F_{d(exit)}$
1	1	1	1.003	1.003
2	1	2	1.002	1.015
3	1	3	1.001	1.010
4	2	1	1.351	1.295
5	2	2	1.045	1.059
6	2	3	1.037	1.050
7	3	1	1.709	1.311
8	3	2	1.147	1.217
9	3	3	1.076	1.029

Table 5.10. *L_9 Orthogonal array and the computed values of delamination factor in drilling experiments of SUPERPAN DÉCOR (melamine coating layer) MDF panel (adapted from Gaitonde et al. 2008, with permission from Elsevier [GAI 08c])*

In the above example, the same number of levels was defined for all the control factors considered. However, the Taguchi design allows us to define different number of levels for each factor. In such situations, mixed level orthogonal arrays need to be selected for experimentation purposes.

The mixed level OA was used to conduct experiments on milling of SONAEPAN MR MDF panels with cemented carbide end mills for optimizing the surface roughness [GAI 08d]. Table 5.11 gives the factor and their chosen levels in milling experiments. Here, 4 levels are defined for feed rate and 2 levels for cutting speed. The mixed L_8 OA and the measured values of surface roughness are summarized in Table 5.12.

Parameter	Code	Unit	Level			
			1	2	3	4
Feed rate (*f*)	A	m/min	0.50	2.75	3.88	5.00
Cutting speed (*v*)	B	m/min	75	452	-	-

Table 5.11. *Factors and levels selected in milling of SONAEPAN MR MDF panel (adapted from Gaitonde et al. 2008, with permission from Walter de Gruyter GmbH & Co. KG [GAI 08d])*

Trial number	Levels of input parameters		Measured surface roughness R_a (μm)
	A	B	
1	1	1	11.39
2	1	2	8.75
3	2	1	13.52
4	2	2	10.71
5	3	1	14.44
6	3	2	11.54
7	4	1	13.57
8	4	2	10.92

Table 5.12. *L_8 orthogonal array and measured surface roughness values in milling of SONAEPAN MR MDF panel (adapted from Gaitonde et al. 2008, with permission from Walter de Gruyter GmbH & Co. KG [GAI 08d)]*

5.4.2. *Signal-to-noise ratio*

The philosophy behind Taguchi robust design is to control the effect of variations caused by noise factors on product quality characteristics rather than controlling the source of the noise itself [PHA 89]. In order to minimize the variations in the quality characteristic or response, Taguchi introduced a method to transform the repetition data to another value, which is a measure of variation present in the scattered response data. This transformation consists of the computation of signal-to-noise (S/N) ratio (η) that consolidates several repetitions into one performance measure, which reflects the amount of variation present [PHA 89]. The maximization of the S/N ratio simultaneously optimizes the response and minimizes the effect of noise factors. For each entry "*i*" in the OA, if the response (*Ob*) is repeated "*n*" times, then the S/N ratio can be computed as follows [PHA 89; ROS 96]:

(i) smaller-the-better type: $\eta_i = -10 log_{10}\{\frac{1}{n}\sum_{i=1}^{n} Ob_i^2\}$ if Ob needs to be minimized;

(ii) larger-the-better type: $\eta_i = -10 log_{10}\{\frac{1}{n}\sum_{i=1}^{n} Ob_i^{-2}\}$ if Ob needs to be maximized.

5.4.3. *Taguchi optimization procedure*

The Taguchi optimization procedure consists of the analysis of means (ANOM) and analysis of variance (ANOVA) on the S/N ratio of OA [PHA 89]. The ANOM is used to identify the optimal factor level combinations and to estimate the main effects of each factor. ANOM is also employed to find the effect of a factor level which is the deviation it causes from the overall mean response. The ANOVA is used to estimate the error variance for the effects and variance of the prediction error. This analysis is performed on S/N ratios to obtain the contribution of each of the factors.

The ANOM consists of determination of the overall mean of η associated with n trials and is given by:

$$m = \frac{1}{n}\sum_{q=1}^{n} \eta_q \qquad [5.12]$$

The effect of a factor level i for a parameter j is defined as:

$$(m)_{i,j} = \frac{1}{l}\sum_{i=1}^{l} (\eta_i)_j \qquad [5.13]$$

The optimum level for a factor is the level that gives the highest value of S/N ratio. The maximization of S/N ratio is determined by ANOM to give the optimum level associated with each factor and is given as:

$$j_{i,opt} = max\{(m)_{i,j}\} \quad \text{for } j = 1, 2, \ldots, k. \qquad [5.14]$$

After determining the optimal levels of control factors, the ANOVA is then performed to estimate the relative significance of each factor [PHA 89].

5.4.4. *Taguchi multi-objective optimization*

The Taguchi optimization technique is suitable for a process with only one objective as the original Taguchi design is designed to optimize single performance characteristics. On the other hand, the application of the Taguchi technique for optimization of a process with multi-objective functions requires modifications. The methodology of Taguchi multi-objective optimizations using "*utility concept*" and "*quality loss function*" are detailed in the following examples.

5.4.4.1. *Taguchi approach with utility concept – a case study*

This example demonstrates the Taguchi multi-objective optimization to determine optimum minimum quantity of lubricant (MQL) and cutting conditions (cutting speed and feed rate) in order to enhance the machinability during turning of brass using K10 carbide tool [GAI 08e]. The Taguchi technique with the utility concept, a multi-response optimization method, has been employed for simultaneous *minimization* of two objectives namely, surface roughness (R_a) and specific cutting force (K_s). The experiments are planned as per Taguchi's L_9 orthogonal array with each experiment performed under different conditions of MQL, cutting speed and feed rate. Table 5.13 presents the factors and their levels and Table 5.14 gives L_9 OA with the responses.

The S/N ratio associated with the responses R_a and K_s are given by:

$$\eta_1 = -10 \log_{10} [R_a^2] \quad [5.15]$$

$$\eta_2 = -10 \log_{10} [K_s^2] \quad [5.16]$$

In the utility concept, the multi-response S/N ratio for each trial in an orthogonal array is computed as [KUM 00]:

$$\eta = w_1 \eta_1 + w_2 \eta_2 \quad [5.17]$$

where w_1 and w_2 are the weighting factors associated with the S/N ratio of each of responses R_a and K_s respectively. A weighting factor of 0.5 for each of the responses is employed. This gives equal priorities to both R_a and K_s for simultaneous minimization. The computed values of S/N ratio for each response and multi-response S/N ratio (η) for each trial in the orthogonal array are illustrated in Table 5.14. Figure 5.5 presents the results of ANOM and Table 5.15 summarizes the ANOVA results. The optimum level combinations of control factors, as seen from Figure 5.5, are 3, 2 and 1. This corresponds to optimal values of quantity of lubricant of 200 ml/hour, cutting speed of 200 m/min and feed rate of 0.05 mm/rev.

Factor	Code	Unit	Level		
			1	2	3
Quantity of lubricant (Q)	A	ml/hr	50	100	200
Cutting speed (v)	B	m/min	100	200	400
Feed rate (f)	C	mm/rev	0.05	0.10	0.15

Table 5.13. *Process parameters and levels selected in turning of brass (adapted from Gaitonde et al. 2008, with permission from Elsevier [GAI 08e])*

Trial no.	Levels of input process parameters			Responses		Calculated S/N ratio (dB)		
	A	B	C	R_a (μm)	K_s (MPa)	η_1	η_2	η
1	1	1	1	0.3	375.1	10.4576	-51.4829	-20.5127
2	1	2	2	0.5	323	6.0206	-50.1841	-22.0817
3	1	3	3	1.1	301.5	-0.8279	-49.5857	-25.2068
4	2	1	2	0.44	307.4	7.1309	-49.7541	-21.3116
5	2	2	3	0.9	286.1	0.9151	-49.1304	-24.1076
6	2	3	1	0.32	374.9	9.8970	-51.4783	-20.7907
7	3	1	3	0.78	304.3	2.1581	-49.6660	-23.7540
8	3	2	1	0.24	343	12.3958	-50.7059	-19.1551
9	3	3	2	0.42	317.9	7.5350	-50.0458	-21.2554

Table 5.14. *L_9 Orthogonal array, responses and the computed values of S/N ratio of multi-performance characteristic in turning of brass (adapted from Gaitonde et al. 2008, with permission from Elsevier [GAI 08e])*

As seen from ANOVA (Table 5.15), the feed rate is the most significant factor contributing around 89%. The error contribution is only 0.55% indicating negligible interaction effects.

Factor	Degrees of freedom	Sum of squares	Mean square	% contribution
A	2	2.2158	1.1079	7.24
B	2	0.7224	0.3612	2.36
C	2	27.4957	13.7478	89.85
Error	2	0.1681	0.0840	0.55
Total	8	30.6020	3.8252	100

Table 5.15. *Summary of ANOVA for multi-performance characteristics in turning of brass (adapted from Gaitonde et al. 2008, with permission from Elsevier [GAI 08e])*

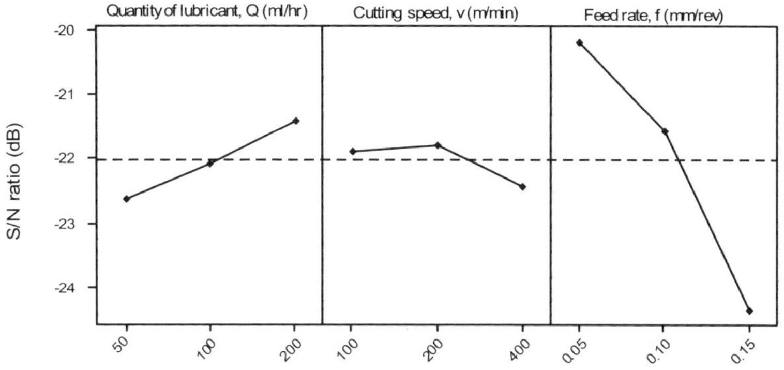

Figure 5.5. *Factor response plot for multiple performance characteristics in turning of brass (adapted from Gaitonde et al. 2008, with permission from Elsevier [GAI 08e])*

5.4.4.2. *Taguchi approach with quality loss concept – a case study*

Consider the multi-objective optimization problem using Taguchi's quality loss function [GAI 08f] for optimizing the machinability during turning of unreinforced polyamide (PA6). The objective is to select suitable tool material (K10 or PCD) and to determine optimal cutting conditions (feed rate and cutting speed) in order to simultaneously minimize the specific cutting force (K_s) and power (P). Three levels are defined for feed rate and cutting speed as given in Table 5.16. The experiments are planned as per L_{18} OA. Table 5.17 summarizes the experimental plan and the two objective functions (responses).

Parameter	Code	Unit	Level		
			1	2	3
Tool material	A	-	K10	PCD	-
Feed rate	B	mm/rev	0.05	0.10	0.15
Cutting speed	C	m/min	50	100	200

Table 5.16. *Process parameters and their levels selected in turning of unreinforced polyamide (PA6) (adapted from Gaitonde et al. 2008, with permission from SAGE [GAI 08f])*

Trial no.	Levels of process parameters			Response	
	A	B	C	P (W)	K_s (MPa)
1	1	1	1	26.35	316.2
2	1	1	2	53.67	322
3	1	1	3	115.7	347.1
4	1	2	1	44.95	269.7
5	1	2	2	90.07	270.2
6	1	2	3	176.4	264.6
7	1	3	1	65.77	263.07
8	1	3	2	130.07	260.13
9	1	3	3	246.27	246.27
10	2	1	1	21.59	259.1
11	2	1	2	40.7	244.2
12	2	1	3	83.57	250.7
13	2	2	1	37.92	227.5
14	2	2	2	73.95	221.85
15	2	2	3	149.8	224.7
16	2	3	1	58.07	232.27
17	2	3	2	104.73	209.47
18	2	3	3	266.37	266.37

Table 5.17. *Orthogonal array and computed values of power and specific cutting force as per experimental layout plan for polyamide turning (adapted from Gaitonde et al. 2008, with permission from SAGE [GAI 08f])*

The quality loss function for i^{th} response at j^{th} trial in an OA for the *smaller-the-better* type characteristic is given by [AME 97, ANT 01]:

$$L_{ij} = y_{ij}^{2} \quad [5.18]$$

where y_{ij} is the ith response in the jth trial. As each response has different units of measurement, it is necessary to normalize the quality loss. The normalized quality loss associated with i^{th} response at jth trial is given by [AME 97, ANT 01]:

$$N_{ij} = \frac{L_{ij}}{\bar{L}_i} \qquad [5.19]$$

where $\bar{L}_i = \frac{1}{n}\sum_{i=1}^{n} L_{ij}$ is the average loss function of ith response due to n trials.

In the weighting method of computing total quality loss function, it is required to assign proper weighting factors to each of the normalized quality loss functions. For "p" number of responses, the total loss function for j^{th} trial can be expressed as [AME 97, ANT 01]:

$$TL_j = \sum_{i=1}^{p} w_i N_{ij} \qquad [5.20]$$

where w_i is the scalar weighting factor for i^{th} response.

The Taguchi loss function for multi-response optimization requires the maximization of multi-response S/N ratio (η), which can be expressed as [AME 97, ANT 01]:

$$\eta_j = -10\log(TL_j) \qquad [5.21]$$

The calculated values of the multi-response S/N ratio (η) are illustrated in Table 5.18. The ANOM is performed to determine the optimal levels of control factors and the response graph is shown in Figure 5.6. The ANOM results [GAI 08f] indicate that a PCD tool with 0.05 mm/rev feed rate and 50 m/min cutting speed is appropriate to simultaneously minimize the specific cutting force and power. Table 5.19 presents the summary of ANOVA for Taguchi optimization in PA6 machining.

Trial no.	Loss function		Normalized loss function		Total loss function (TL_j)	Signal-to-noise ratio (η), dB
	L_{1j}	L_{2j}	N_{1j}	N_{2j}		
1	694.3225	99982.44	0.047367	1.442879	0.745123	1.277722
2	2880.469	103684	0.196507	1.496297	0.846402	0.724234
3	13386.49	120478.4	0.913231	1.738662	1.325947	-1.22526
4	2020.503	72738.09	0.137839	1.049707	0.593773	2.263795
5	8112.605	73008.04	0.553445	1.053602	0.803524	0.950013
6	31116.96	70013.16	2.122811	1.010382	1.566597	-1.94957
7	4325.693	69205.82	0.2951	0.998731	0.646916	1.891522
8	16918.2	67667.62	1.154166	0.976533	1.06535	-0.27492
9	60648.91	60648.91	4.137492	0.875244	2.506368	-3.99045
10	466.1281	67132.81	0.031799	0.968815	0.500307	3.007632
11	1656.49	59633.64	0.113006	0.860592	0.486799	3.126501
12	6983.945	62850.49	0.476447	0.907015	0.691731	1.600625
13	1437.926	51756.25	0.098096	0.746911	0.422503	3.741698
14	5468.603	49217.42	0.37307	0.710272	0.541671	2.662642
15	22440.04	50490.09	1.530868	0.728639	1.129753	-0.52984
16	3372.125	53949.35	0.230048	0.77856	0.504304	2.973076
17	10968.37	43877.68	0.748267	0.633213	0.69074	1.606856
18	70952.98	70952.98	4.840439	1.023945	2.932192	-4.67192

Table 5.18. *Tabulation of computed values of multi-response S/N ratio for PA6 polyamide (adapted from Gaitonde et al. 2008, with permission from SAGE [GAI 08f])*

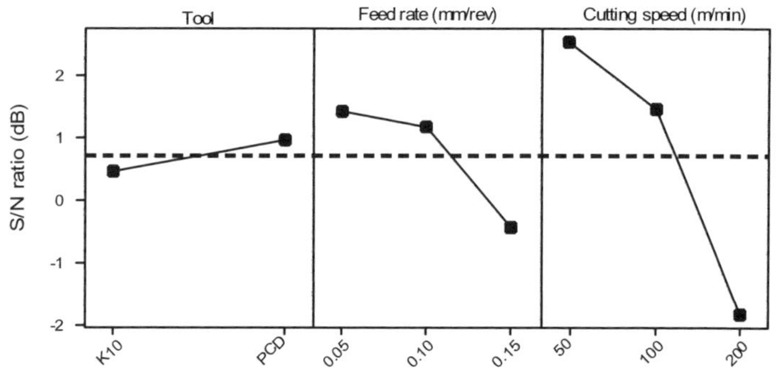

Figure 5.6. *Main effect plot of S/N ratio for machinability in PA6 polyamide machining (adapted from Gaitonde et al. 2008, with permission from SAGE [GAI 08f])*

Parameter	Degrees of freedom	Sum of squares	Mean square	Percent contribution
Tool material	1	10.6571	10.6571	10.72
Feed rate	2	11.9241	5.9620	11.99
Cutting speed	2	60.8364	30.4182	61.20
Error	12	15.9905	1.3325	16.09
Total	17	99.4081	5.8475	100

Table 5.19. *Summary of ANOVA for PA6 polyamide machining (adapted from Gaitonde et al. 2008, with permission from SAGE [GAI 08f])*

5.5. Summary

This chapter addressed the importance of DOE in machining process. The experimental layout plan as per DOE gives a systematic way to conduct the experiments more effectively at lower cost. The selection of proper DOE is governed by many factors such as purpose of conducting machining experiments and the number of factors and their level settings.

The details of experimental layout plan as per FFD, CCD, BBD and OA were discussed using the machining examples like turning of CFRP composite, milling and drilling of MDF panels and drilling of AISI 316/316L stainless steel materials. The DOE based on OA is always advantageous as compared to other designs since OA allows direct optimization without the need for the development of a mathematical model.

The mathematical model development using RSM is based on regression analysis and is quite a useful methodology. The RSM-based models indicate the main and interaction effects of various control factors on the quality characteristic.

The main strategy of Taguchi robust design is to make the process response less sensitive to noise factors. The Taguchi optimization employs the experiments based on OA and analysis performed on the S/N ratio. The multi-objective Taguchi optimization using “utility” and “quality loss” concepts are illustrated with machining examples.

5.6. Acknowledgement

The authors would like to thank Springer, Elsevier, Walter de Gruyter GmbH & Co. KG and Sage Publications for granting them permission to reuse the published materials.

5.7. References

[AME 97] AMES A. E., MATTUCCI N., MACDONALD S., SZONYI G., HAWKINS D. M., "Quality loss functions for optimization across multiple response surfaces", *Journal of Quality Technology*, vol. 29, 1997, p. 339-346.

[ANT 01] ANTONY J., "Simultaneous optimisation of multiple quality characteristics in manufacturing processes using Taguchi's quality loss function", *International Journal of Advanced Manufacturing Technology*, vol. 17, p. 134-138.

[COC 92] COCHRAN W. G., COX G. M., *Experimental Designs*, John Wiley, New York, 1992.

[GAI 08a] GAITONDE V. N., KARNIK S. R., RUBIO J. C., CORREIA A. E., ABRAO A. M., DAVIM J. P., "Analysis of parametric influence on delamination in high speed drilling of carbon fiber reinforced plastic composites", *Journal of Materials Processing Technology*, vol. 203 no. 1-3, 2008, p. 431-438.

[GAI 08b] GAITONDE V. N., KARNIK S. R., SIDDESWARAPPA B., ACHYUTHA B. T. "Integrating Box-Behnken design with genetic algorithm to determine the optimal parametric combination for minimizing burr size in drilling of AISI 316L stainless steel", *International Journal of Advanced Manufacturing Technology*, vol. 37 no. 3-4, 2008, p. 230-240.

[GAI 08c] GAITONDE V. N., KARNIK S. R., DAVIM J. P., "Taguchi multiple-performance characteristics optimization in drilling of medium density fibreboard (MDF) to minimize delamination using utility concept", *Journal of Materials Processing Technology*, vol. 196 no. 1-3, 2008, p. 73-78.

[GAI 08d] GAITONDE V. N., KARNIK S. R., DAVIM J. P., "Prediction and optimization of surface roughness in milling of medium density fiberboard (MDF) based on Taguchi orthogonal array of experiments", *Holzforschung*, vol. 62 no. 2, 2008, p. 209-214.

[GAI 08e] GAITONDE V. N., KARNIK S. R., DAVIM J. P., "Selection of optimal MQL and cutting conditions for enhancing machinability in turning of brass", *Journal of Materials Processing Technology*, vol. 204 no. 1-3, 2008, p. 459-464.

[GAI 08f] GAITONDE V. N., KARNIK S. R., FRANCISCO M., DAVIM J. P., "Taguchi approach for achieving better machinability in unreinforced and reinforced polyamides", *Journal of Reinforced Plastics and Composites*, vol. 27 no. 9, 2008, p. 909-924.

[KUM 00] KUMAR P., BARUA P. B., GAINDHAR J. L., "Quality optimization (multi-characteristic) through Taguchi technique and utility concept", *Quality and Reliability Engineering International*, vol. 16 no. 6, 2000, p. 475-485.

[MIN 06] MINITAB INC., Minitab user manual, version 13, Quality Plaza, 1829 Pine Hall Road, State College, PA 16801-3008, USA, 2006.

[MON 01] MONTGOMERY D. C., *Design and Analysis of Experiments*, John Wiley, New York, 1996.

[PHA 89] PHADKE M. S., *Quality Engineering using Robust Design*, Prentice Hall, Englewood Cliffs, New Jersey, 1989.

[ROS 96] ROSS P. J., *Taguchi Techniques for Quality Engineering*, McGraw-Hill, Singapore, 1996.

Chapter 6

Single and Multi-objective Optimization Methods

6.1. Machining optimization

Machining optimization is the systematic and scientific approach for solving problems associated with engineering decision making in selecting the best production method, process parameters or operational conditions. The ultimate goal in such problems is to either minimize the effort required or maximize the desired benefits. Efforts required and benefits desired in a machining situation can be expressed according to the decision variables. Machining optimization can be defined as the process of finding such decision variables that give the maximum or minimum value of one or more objective functions subject to some resources, capacities or process constraints. In the presence of one objective function, machining optimization becomes a single objective optimization problem. However, when there is more than one objective function with a conflicting or competing relationship, machining optimization becomes a bi-objective or multi-objective optimization problem.

While many conventional optimization techniques such as linear programming exist to solve such problems, the intelligent techniques such as

---

Chapter written by Tugrul ÖZEL.

genetic algorithms, simulated annealing, particle swarm optimization, tabu search and others offer more speedy and robust solutions. Detailed explanations about such techniques can be found in several other books [SAR 06].

We will focus on one particular machining optimization problem in this section. In a globally competitive manufacturing environment, simultaneously increasing productivity, while improving and maintaining high quality surfaces in the finish turning of hardened parts (hard turning), is required to gain a competitive advantage for manufacturers. Hard turning processes demand the use of advanced tools with specially prepared cutting edges as reported in previous literature. König *et al.* [KÖN 84] presented the advantages of using hard turning over grinding and its potential to control surface integrity by optimizing tool geometry and machining parameters. Matsumoto *et al.* [MAT 99] addressed the issue of obtaining favorable surface integrity in hard turning processes. It is also evident from a large number of other experimental works that the tool geometry and selected machining parameters have complex relations with cutting forces, tool life, surface roughness and integrity of the finished surfaces. Chou *et al.* [CHO 02] showed the influence of using various Cubic Boron Nitrite (CBN) contents in the cutting tools on tool wear and surface integrity. Thiele *et al.* [THI 00] have shown that cutting edge geometry has great influence on the residual stresses induced by hard turning. Özel *et al.* [ÖZE 05a] have identified the factors influencing cutting forces and surface roughness. This complex nature of parameter relations makes it difficult to generate explicit analytical models for hard turning processes. On the other hand, according to a review by Dimla *et al.* [DIM 97], the relations between the machining parameters including tool geometry and the performance measure of interest can be obtained with non-linear models by training neural networks using experimental data as reported by many researchers. Özel and Nadgir [ÖZE 02] applied neural network modeling to predict tool wear in different edge geometry tools in hard turning. It is also highly complicated to predict surface roughness generated by machining processes as claimed by Benardos and Vosniakos [BEN 03]. Özel and Karpat [ÖZE 05b] have also shown that neural network modeling is a viable approach for predicting both surface roughness and tool wear for honed and chamfered cutting tools.

The general formulation of multi-objective optimization problems can be written in the following form:

Minimize (or maximize) $\mathbf{f}(\mathbf{x})=\{f_1(\mathbf{x}), f_2(\mathbf{x}), f_3(\mathbf{x})..., f_k(\mathbf{x})\}$
subject to $g_j(\mathbf{x}) \leq b_j$ for $j = 1,2,...,m$
and $h_j(\mathbf{x}) = b_j$ for $j = m+1,...,m+p$

In this formulation; $f_i(\mathbf{x})$ denotes the ith objective function, $g_i(\mathbf{x})$ and $h_i(\mathbf{x})$ indicate inequality and equality type of constraints and the decision variables (machining parameters and tool geometry) are shown with the vector **x**, where $\mathbf{x} = (x_1, x_2, ..., x_n) \in \Re^N$. The ultimate goal is simultaneous minimization or maximization of given objective functions. When, as in most cases, some of the objective functions conflict with each other, there is no exact solution but many alternative solutions. This family of potential solutions cannot improve all the objective functions simultaneously. This is called Pareto optimality [PAR 71] and the definition is given below.

Definition 1. Pareto optimal: consider a point $\mathbf{x}^*$ in the feasible solution space, **X**, $\mathbf{x}^* \in \mathbf{X}$. The point (a set of decision variables) is Pareto optimal if and only if there does not exist another point, $\mathbf{x} \in \mathbf{X}$, that satisfies $\mathbf{f}(\mathbf{x}) < \mathbf{f}(\mathbf{x}^*)$ and $f_i(\mathbf{x}) < f_i(\mathbf{x}^*)$ for at least one function.

In other words, this definition states that, for a minimization problem, there is no other point which can cause a decrease in one objective function value without causing a simultaneous increase in at least one of the other objective function values.

Definition 2. Dominated and non-dominated points: a vector of objective functions, $\mathbf{f}(\mathbf{x}^*)$, is non-dominated if and only if there does not exist another vector, $\mathbf{f}(\mathbf{x})$, that satisfies $\mathbf{f}(\mathbf{x}) \leq \mathbf{f}(\mathbf{x}^*)$ with at least one $f_i(\mathbf{x}) < f_i(\mathbf{x}^*)$. Otherwise, $\mathbf{f}(\mathbf{x}^*)$ is dominated.

Definition 3. Pareto front: the set $\mathbf{X}^* = \{\mathbf{x}_1^*, \mathbf{x}_2^*, ..., \mathbf{x}_n^*\}$, which is composed of all the non-dominated Pareto optimal solutions that comprise the Pareto front of non-dominated solutions.

There are numerous methods used to solve multiple objective optimization problems. The most common method is to combine all objectives into a single objective function through the use of "weights" or utility functions and solve for a single solution as reported by Marler and Arora [MAR 04]. The weighted-sum method is applied for multi-parameter turning optimization using neural network modeling and particle swarm optimization in Karpat and Özel [KAR 05]. The combined objectives approach yields a single solution that can be readily implemented, but this solution largely depends on numerical weights or utility functions that are often difficult to select, and are often selected somewhat arbitrarily. The Pareto optimal non-dominated solution set avoids this problem and may provide numerous prospective solutions (sets of machining parameters and tool geometry) for the decision maker (manufacturer) during process planning for hard turning processes.

6.2. Intelligent optimization techniques

In obtaining optimal solutions for these multi-objective problems, a number of heuristic algorithms including genetic algorithms (GA), evolutionary algorithms (EA) and simulated annealing (SA), as well as a recently developed optimization algorithm called particle swarm optimization (PSO) are used. These methods are summarized by Zitzler *et al.* [ZIT 03]. The PSO is a population-based search algorithm that was inspired by the social behavior of flocks of birds as originally developed by Eberhart and Kennedy [EBE 95]. The PSO operates on a population of potential solutions. Conventional optimization techniques, such as gradient-based methods, do not function effectively solving multiple objective functions without reducing the objectives into a single function and handling it accordingly. On the other hand, the PSO algorithms have some drawbacks such as exploration and exploitation while searching the solution space [SHI 98]. Many researchers such as Coello and Lechuga [COE 02] find the PSO to be an efficient alternative over other search algorithms especially when dealing with multi-objective optimization problems.

Since conventional gradient-based optimization techniques are not suitable for solving multi-objective optimization problems, meta-heuristic search algorithms are often employed in these cases. Stochastic and population-based search algorithms are suitable for finding Pareto optimal solutions for multi-objective functions. Many GA-based approaches have been proposed to solve multi-objective optimization problems for machining processes, e.g. Cus and Balic [CUS 03], and Saravanan *et al.* [SAR 02]. The PSO has reportedly performed as well as GAs on many test problems [COE 02], and it is an efficient and general method to locate the Pareto front of the multi-objective optimization problems. The PSO has been applied successfully to predicting surface roughness in end milling by El-Mounayri *et al.* [ELM 03] and optimizing cutting conditions for NC end milling by Tandon *et al.* [TAN 02]. More recently, Karpat and Özel [KAR 05] applied PSO to hard turning optimization where tool wear and surface roughness is modeled by neural network modeling and machining parameter optimization is performed by considering tool wear and surface roughness requirements for a given cutting distance. Tool wear and surface roughness requirements were combined in a single objective function and a single optimum cutting condition was obtained by using a conventional PSO algorithm.

In this section, different from the work presented in Karpat and Özel [KAR 05], objective functions will be treated separately and a group of optimal cutting conditions will be calculated using dynamic neighborhood particle swarm optimization (DN-PSO). DN-PSO was originally proposed by Hu and Eberhart [HU 02] and utilized to solve optimization problems with multiple objective functions. Thus, the objective of this analysis is to obtain a group of optimal process parameters which minimize surface roughness values and maximize the productivity at the same time, or maximize tool life while maximizing material removal rate, or minimize tensile stresses on the surface while maximizing compressive stresses underneath the machined surface for most optimal surface integrity.

It is widely reported that the PSO algorithm is very easy to implement and has fewer parameters to adjust when compared to other evolutionary algorithms. The information sharing mechanism among the particles in PSO is significantly different from the information sharing among the chromosomes in GAs. In GAs, the entire group moves towards an optimal solution area. However, in PSO only the global best or local best solution is reported to the other particles in a swarm. Therefore, evolution only looks for the best solution and the swarm tends to converge to the best solution quickly and efficiently.

PSO is a population-based stochastic optimization technique developed by Eberhart and Kennedy [EBE 95] inspired by the social behavior of flocks of bird. The PSO is initialized with a population of random solutions and this initial population evolves over generations to find optima. In PSO, different from genetic algorithms, each particle in the population has a velocity, which enables it to fly through the problem space. Therefore, each particle is represented by a position and a velocity vector. Dimensions of position and velocity vectors are defined by the number of decision variables in the optimization problem. Modification of the position of a particle is performed by using its previous position information and its current velocity. According to the value of the objective function, each particle knows its best position ever (personal best, *pbest*) and the best position achieved in the group (group best, *gbest*) among all personal bests. For a single objective problem, the result of the optimization problem will be the position vector of *gbest* at final iteration. These principles can be formulated as:

$$v_i^{k+1} = wv_i^k + c_1 rand_1 \left(pbest_i - x_i^k \right) + c_2 rand_2 \left(gbest_i - x_i^k \right) \qquad [6.1]$$

where

v_i^k: velocity of agent i at iteration k
x_i^k: current position of agent i at iteration k
$pbest_i$: personal best of agent i
$gbest$: best position in the neighborhood
$rand$: random number between 0 and 1
w: weighting function
c_j: learning rate j=1,2

$$x_i^{k+1} = x_i^k + v_i^{k+1} \qquad [6.2]$$

The first term on the right hand side of equation [6.1] is the previous velocity of the particle. Weighting function w is set at a large value at the beginning of the search and decreased to a smaller value over the iterations to confine the search in a smaller region in later iterations, or it could be selected randomly. The second and third terms are used to change the velocity of the particle according to *pbest* and *gbest* values. The random numbers used in the velocity update step give the PSO a stochastic behavior. The iterative approach of PSO can be described as follows.

Step 1: first a population size is specified. Initial positions and velocities of agents are generated randomly. Then, objective function values for each agent (or particle) are calculated. For the first iteration, *pbest* is set as the current position of each particle. The *pbest* with best objective function value among the agents is set as *gbest* and this value is stored.

Step 2: in the next iteration, the new position of the agents (particles) in the solution space is determined by using equations [6.1] and [6.2].

Therefore, the particles begin to move towards the particle with best objective function value, *gbest*.

Step 3: the objective function value is calculated for new positions of each particle. If an agent achieves a better position, the *pbest* value is replaced by the current value. As in Step 1, *gbest* value is selected among *pbest* values. If the new *gbest* value is better than the previous *gbest* value, the *gbest* value is replaced by the current *gbest* value and stored.

Step 4: steps 1, 2 and 3 are repeated until a predetermined iteration number is reached.

The velocity update step in the PSO is stochastic due to the random numbers generated, which may cause an uncontrolled increase in velocity and therefore instability in the search algorithm. In order to prevent this, velocities are limited to the dynamic range of the particle on each dimension [EBE 95].

The formulation of the PSO given with equations [6.1] and [6.2] corresponds to the global version of the PSO. In addition, a local version of the PSO algorithm also exists. In the local version, particles have information about only of their own and their nearest neighbor best (*lbest*) [EBE 95]. *gbest* is then replaced by *lbest* in the algorithm.

The version and variations of PSO given above are not suitable for solving multi-objective optimization problems since there is no absolute global minimum (or maximum). Therefore, the algorithm needs some modifications to locate the Pareto front in multi-objective optimization problems. Hu and Eberhart [HU 02] proposed DN-PSO for this purpose. In this approach, instead of defining one global best for the whole population, neighborhoods are defined for each particle and local *gbest*s are found within these neighborhoods.

Figure 6.1 shows a 2D objective function space in a min-min problem. The Pareto front is the boundary of the objective value region, which is the lower left side of the objective function space for min-min problems. The objective of the proposed algorithm is to drop those solutions onto the boundary line indicated by a solid line.

For this purpose, the first objective function is fixed to define neighborhoods and the second objective function is used in optimization. According to the DN-PSO algorithm; (1) the distances between the current particle and other particles are calculated in terms of the first objective function; (2) based on these distances, the nearest m (neighborhood size) particles are found; (3) the local best particle among neighbors is selected in terms of the second objective function as illustrated in Figure 6.2.

In this analysis, the neighborhood size is selected as m=2. As an example, for a particle whose objective function values correspond to point 6 in the objective function space, the nearest neighbors in terms of the first objective function are points 5 and 7. Since point 5 has a lower objective function value in terms of the second objective function, point 5 is selected as a local *gbest* for point 6. Similarly, point 3 is selected as the local *gbest* for point 1. This procedure is repeated for all particles and local *gbest*s for each particle are found. Since after the each iteration the positions of the particles, and therefore their objective values, change, neighborhoods are defined dynamically at each step.

Another important term *pbest*, which is the best position in the memory of a particle, is updated according to the Pareto dominance principle. The new *pbest* should simultaneously improve both objective functions as shown in Figure 6.3.

In the PSO, only two parameters need to be defined, weighting function w and learning rates c_1 and c_2. The values of these parameters are adapted from Hu and Eberhart [HU 02] where weight w was chosen as (0.5 + (Rand/2)) and learning rates were taken as 1.49445. These parameters appear to work well with many benchmark multi-objective optimization problems [HU 02].

The formulation given above is only valid for two-objective optimization problems. For cases where there are more than two objective functions, different algorithms have been proposed by Coello and Lechuga [COE 02], Fieldsend and Singh [FIE 02], Mostaghim and Teich [MOS 03]. Since the PSO and its variations are relatively new, development of better methodologies is an ongoing research area.

6.3. Case studies in machining optimization

Hard turning machining exhibits a unique behavior, which is different from conventional turning operations. Application of hard turning technology can be improved by utilizing advanced optimization algorithms which help manufacturers to make educated decisions when faced with multiple objectives to be satisfied. Finish hard turning, using Cubic Boron Nitride (CBN) tools, allows manufacturers to simplify their processes and still achieve the desired surface roughness, which can compete with grinding operations.

Surface roughness is mainly a result of process parameters such as tool geometry (i.e. nose radius, edge geometry, rake angle, etc.) and cutting conditions (feed rate, cutting speed, depth of cut, etc.). In finish hard turning, tool wear becomes an additional parameter affecting the surface quality of finished parts. The performance of CBN cutting tools is highly dependent on cutting conditions such as cutting speed, feed, feed-rate and depth of cut. Cutting speed and depth of cut have a particularly significant influence on tool life.

It has been observed that decreasing feed rate helps obtain a good surface finish but increases machining time. High cutting speeds may help reduce the surface roughness, but since tool life at high cutting speeds is just a few minutes this solution is not applicable. In some cases surface roughness is improved with increasing tool wear. Therefore, attention should be paid to the relationship between tool wear and surface roughness. It is therefore crucial to obtain a group of optimum conditions, which may serve different purposes under different circumstances.

6.3.1. *Case study 1: minimizing surface roughness and minimizing machining time*

It has been shown that the actual surface roughness obtained in experiments does not match the theoretical values due to adhesion, ploughing and geometric effects, which cannot easily be modeled with analytical models. To overcome this problem, neural networks, which are known to have a great capacity for capturing actual process information, are designed with the data obtained from surface roughness and tool wear experiments.

Tool wear and surface roughness behavior of AISI H13 steel machined with honed and chamfered CBN tools (honed inserts with a radius of 0.01 mm and chamfered inserts with 0.1 mm chamfer land and 20° chamfer angle) are observed [ÖZE 05b]. The experimental results presented by Özel and Karpat [ÖZE 05b] were obtained by longitudinal turning of hardened AISI H13 steel bars with 54.7 HRC that was conducted on a rigid, high-precision, production type CNC lathe (Romi Centur 35E) at a constant depth of cut at 0.254 mm. The bar work pieces were held in the machine with a collet to minimize run-out and maximize rigidity. The length of cut for each test was 25.4 mm in the axial direction. Due to availability constraints, each insert was used for one factor-level combination, which consisted of 16 replications (a total of three honed and three chamfer inserts were available). In this manner each edge preparation was subject to the same number of tests and the same axial length of cut. Finally, surface roughness and tool wear measurements were conducted after machining axial cutting length of 25.4 mm (1 inch) up to 406.4 mm (16 inches) during each factor-level combination. The surface roughness was measured with a Taylor-Habson Surtronic 3+ profilometer and Mitutoyo SJ-digital surface analyzer, using a trace length of 4.8 mm, a cut-off length of 0.8 mm, and an M1 band-pass filter [ÖZE 05b]. The surface roughness values were recorded at eight equally spaced locations around the circumference every 25.4 mm from the edge of the specimen to obtain statistically meaningful data for each factor level combination [ÖZE 05b]. CBN inserts were examined using a tool-maker microscope to measure flank wear depth and detect undesirable features on the edge of the cutting tool by interrupting the finish hard turning process [ÖZE 05b]. The factor levels and measured tool wear after 406.4 mm of cutting as well as average surface roughness values for honed and chamfered tools are given in Table 6.1.

Honed Inserts					Chamfered Inserts				
Cutting Speed (m/min)	Feed Rate (mm/rev)	Axial CuttingLength (mm)	Tool Wear (mm)	Surface. Roughness (μm)	Cutting Speed (m/min)	Feed Rate (mm/rev)	Axial Cutting Length (mm)	Tool Wear (mm)	Surface Roughness (μm)
100	0.1	406.4	0.13	0.38	100	0.1	406.4	0.2	0.5
200	0.1	406.4	0.16	0.33	200	0.1	406.4	0.25	0.54
100	0.2	406.4	0.13	0.42	100	0.2	406.4	0.25	1.25
200	0.2	406.4	0.13	1	200	0.2	406.4	0.15	1.15
200	0.05	406.4	0.15	0.25	N/A				

Table 6.1. *Summary of the experimental results reported by Özel and Karpat [ÖZE 05b]*

The neural network, designed for the first case study, takes cutting speed, feed rate and cutting length as inputs and predicts tool wear and surface roughness along the work piece cutting length. For the first neural network, which predicts tool wear 3-6-1, and for the second neural network, which predicts surface roughness 4-6-1, neural network architecture is selected. The details of the selection process are given in Özel and Karpat [OZE 05b].

The following optimization problem can be defined for chamfered and honed tools separately as below and used in multi-objective particle swarm optimizers in SINNS architecture. Decision variables cutting speed V and feed rate f are constrained within the ranges of the experiments.

$$
\begin{aligned}
&f_1 = \text{minimize } (\text{Surface roughness})\\
&f_2 = \text{minimize } \left(\text{Machining time} = \frac{L}{f \times N}\right)\\
&\text{subject to} \qquad\qquad [6.5]\\
&100 \text{ m/min} \le \text{V} \le 200 \text{ m/min}\\
&0.05 \text{ mm/rev} \le \text{f} \le 0.2 \text{ mm/rev} (\text{Honed inserts})\\
&0.1 \text{ mm/rev} \le \text{f} \le 0.2 \text{ mm/rev} (\text{Chamfered inserts})
\end{aligned}
$$

In order to handle constraints, a simple modification to the particle swarm optimization algorithm is sufficient. The additional rules that should be implemented in a DN-PSO algorithm are: (1) particles should be initiated within a feasible region; (2) when updating memories, only the particles within a feasible region should be kept in the memory. The DN-PSO optimization algorithm combined with constraints was converted into a code in Matlab software.

The results obtained for chamfered and honed tools are shown in Figure 6.7 and 6.8. The DN-PSO algorithm was initiated with 100 particles. After 100 iterations, 77 non-dominated solutions for honed and 73 non-dominated solutions for chamfered are found after 50 iterations. Total computation time is less than 20 minutes with a Pentium IV 1.6 GHz processor. The performance of the proposed SINNS methodology is dependent upon the number of iterations and the population size of the particles in the DN-PSO. The usage of a cascaded neural network system (see Figure 6.6) resulted in larger computation time in this case study.

Since the selection of feed rate influences surface roughness and machining time conversely, minimization of surface roughness and minimization of machining time are contradicting objectives. In order to obtain a good surface finish, feed rate should be reduced, which then increases the machining time. Therefore, a compromise between surface roughness and machining time should be made. According to some candidate solutions listed in Table 6.2, machining time can be reduced more than one minute with a 0.1 micron sacrifice in surface roughness estimation by setting the cutting conditions to *V=200 m/min* and *f=0.125 mm/rev* instead of *V=100 m/min* and *f=0.1 mm/rev*.

As for the honed edge, since lower feed rate experiments were performed during experimentation, better surface roughness values were obtained. Similarly, at the cost of increasing surface roughness slightly shorter machining times can be obtained by adopting the optimum cutting conditions given in Table 6.2. Figures 6.9 and 6.10 demonstrate an increasing trend in material removal rate, an important productivity indicator.

Some of the optimal cutting speed, feed rate pairs from the Pareto fronts with corresponding surface roughness and machining times are shown in Table 6.2.

Honed Inserts				Chamfered Inserts			
Cutting Speed (m/min)	Feed Rate (mm/rev)	Mach. Time (min.)	Surface Roughness (μm)	Cutting Speed (m/min)	Feed Rate (mm/rev)	Mach. Time (min.)	Surface Roughness (μm)
200	0.05	1.58	0.26	100	0.1	1.99	0.61
197.7	0.094	1.07	0.31	200	0.125	0.79	0.70
184.8	0.12	0.84	0.43	192.8	0.147	0.70	0.83
163.4	0.18	0.64	0.74	200	0.175	0.57	1.05
178.7	0.2	0.55	0.93	200	0.2	0.49	1.21

Table 6.2. *Some selected optimal machining parameters for case study 1*

6.3.2. *Case study 2: maximizing tool life and maximizing material removal rate*

As mentioned earlier, wear behavior of CBN tools are different from conventional tools. Thus, the tool life expressions developed for conventional tools does not apply to CBN tools [CHO 02]. In order to be competitive in today's market conditions, manufacturers should find the most economical way to manufacture their product, which can be possible by selecting the optimum machining conditions.

In a recent work by Mamalis *et al.* [MAM 05], a new tool life equation for CBN tools when machining 100Cr6 steel (62 HRC) was proposed (equation [6.6]). This proposed tool life equation, which is valid in the whole cutting speed range for a given feed rate and depth of cut, has two extreme values.

$$T = \frac{A}{V^3 + B \times V^2 + C \times V} \qquad [6.6]$$

Coefficients *A, B,* and *C* of equation [6.6] were calculated under different cutting conditions and the results are in good agreement with experimental values [MAM 05]. It was also shown that different feed rate and depth of cut values change the location of the two extreme values of the tool life curve.

In this analysis, the experimental data given in [MAM 05] is used to design a neural network model which predicts tool life according to given cutting speed, feed rate and depth of cut values. A neural network with two hidden layers and sigmoid transfer functions with a network topology of 3-4-2-1 is chosen and trained with Bayesian regularization. The same procedure is adopted in selecting optimal network topology and using 5% error criteria. The experimental values obtained at a feed rate of 0.075 mm/rev and depth of cut of 0.1 mm are excluded from the data set during training. The comparison of predicted and actual data together with other experimental values is shown in Figure 6.11. As in the previous case study, the neural network model is inserted in SINNS to obtain the Pareto front between tool life and material removal rate. Since it is desirable to maximize both tool life and material removal rate, the necessary modifications in the DN-PSO algorithm should be performed.

The optimization problem can be defined as:

$$\begin{aligned}
&f_1 = \text{maximize } (\text{Tool life}) \\
&f_2 = \text{maximize } (\text{Material removal rate } = V \times f \times d) \\
&\text{subject to} \\
&11 \text{ m/min} \le V \le 120 \text{ m/min} \\
&0.025 \text{ mm/rev} \le f \le 0.125 \text{ mm/rev} \\
&0.05 \text{ mm} \le d \le 0.25 \text{ mm}
\end{aligned} \qquad [6.7]$$

where V is the cutting speed, f is the feed rate and d is the depth of cut. Again, the search is initiated with 100 particles and 90 non-dominated solutions are found after 50 iterations. The computation time was less than 10 minutes in this case. The Pareto front can be seen in Figure 6.12.

Some of the optimal cutting conditions are listed in Table 6.3.

Feed Rate (mm/rev)	Depth of Cut (mm)	Cutting Speed (m/min)	Tool Life (min.)	MRR (mm^3/min)
0.025975	0.11823	11	349.8	0.033781
0.040953	0.21028	49.333	216.47	0.42484
0.072226	0.24096	40.831	194.85	0.71061
0.095481	0.24983	40.396	178.26	0.96361
0.125	0.25	55.466	108.57	1.7333
0.125	0.25	82.985	12.297	2.5933
0.125	0.25	120	5.5151	3.75

Table 6.3. *Some selected optimal machining parameters for case study 2*

6.3.3. *Case study 3: minimizing machining induced stress and minimizing surface roughness*

Machining induced residual stresses, the stresses which remain in a machined part after manufacturing, on the machined surface and subsurface strongly affects the fatigue life of a component. Although some analytical models make use of slip-line theory [JAC 00] or finite element modeling [YAN 02] to predict residual stresses, the majority of the work has been performed experimentally [THI 00; MIT 98; HUA 04; JAC 02; LIU 04; DAH 04]. In their work, Mittal and Liu [MIT 98] focused on the influence of hard turning on residual stresses. They first showed that machining conditions such as cutting speed and feed rate affect the residual stress, and by using experimental observations they adjusted cutting parameters to maximize fatigue life in rolling contact. They concluded that the difference in fatigue life can be 47 times between best and worst machining conditions [MIT 98].

The effects of tool edge geometry, work piece hardness, work piece material and tool wear have also been investigated by various researchers [THI 00; MIT 98; HUA 04; JAC 02; LIU 04; DAH 04]. It has been shown that high work piece hardness, large hone radius and high feed rate results in higher compressive residual stresses in the machined subsurface [HUA 04]. With the increasing tool wear, the residual tensile stresses on the surface increases and the residual compressive stresses on the subsurface also increase remarkably [LIU 04]. On the other hand, the depth of cut has been found to have no effect on residual stresses [LIU 04].

Jacobson *et al.* [JAC 02] performed machining experiments on heat-treated bainite steel with 58 HRC. They selected bainite steel to show its favorable properties in terms of residual stresses and proposed it as an alternative material for AISI 52100 as bearing steel, which is commonly used in hard turning. They machined so-called B8 test rings by using a CBN tool -6° rake angle with 0.1 mm/rev feed and 0.1 depth of cut with a cutting speed ranging from 50 m/min to 999 m/min. The residual stresses in feed and speed direction and surface roughness values are then measured. The measured surface roughness values with respect to cutting speed and residual stresses in feed direction are given in Figures 6.13 and 6.14 as obtained from [JAC 02].

It can be seen from the experimental results reported by Jacobson *et al.* [JAC 02] that the best surface roughness value of 1.7 μm is obtained at cutting speeds around 170 m/min as shown in Figure 6.13, the highest compressive residual stress value of about 225 MPa on the surface (where depth is zero) is obtained at cutting speed of 290 m/min, and the highest minimum subsurface compressive residual stress value of about 680 MPa is obtained at a cutting speed of 230 m/min, as shown in Figure 6.14.

As stated by many researchers, customized residual stress distributions can be obtained by optimizing machining parameters. For this purpose, as in previous case studies, one neural network model for residual stresses and another for surface roughness are designed and cascaded as shown in Figure 6.15. This neural network model takes cutting speed and depth below the surface as inputs and produces residual stress distributions while using cutting speed to predict surface roughness. A 1-3-2-1 neural network model with two hidden layers for surface roughness and a 2-6-4-1 neural network model with two hidden layers are designed and integrated into the SINNS architecture. In the selection of optimal neural network topology, the same procedure as given in previous cases was adopted.

In this case, the optimization problem is written as:

f_1 = minimize (tensile residual stress on the surface)

f_2 = minimize (surface roughness)

subject to

50 m/min $\leq$ V $\leq$ 999 m/min

The Pareto front results for this case study given in Figures 6.15 and 6.16 showed that for the best (minimum) surface roughness values and minimum residual stress values on the surface can be obtained for the optimum cutting speed range between 150 m/min and 250 m/min. Some selected optimal cutting speed parameters are given in Table 6.4. In this case, population size of 100 particles is used and 65 non-dominated solutions on the Pareto front are found after 50 iterations. The computation time was less than 10 minutes.

Cutting Speed (m/min)	Surface Roughness (micron)	Surface Residual Stress (MPa)	Subsurface Residual Stress (MPa)
158.86	1.7364	-134.88	640.04
225.15	1.9407	-147.13	-630.88
230.63	1.9855	-152.6	-623.75
240.63	2.0789	-164.75	-608.36
245	2.1238	-170.81	-600.89
249.43	2.1712	-177.29	-593.02

Table 6.4. *Some selected optimal cutting speed parameters for case study 3*

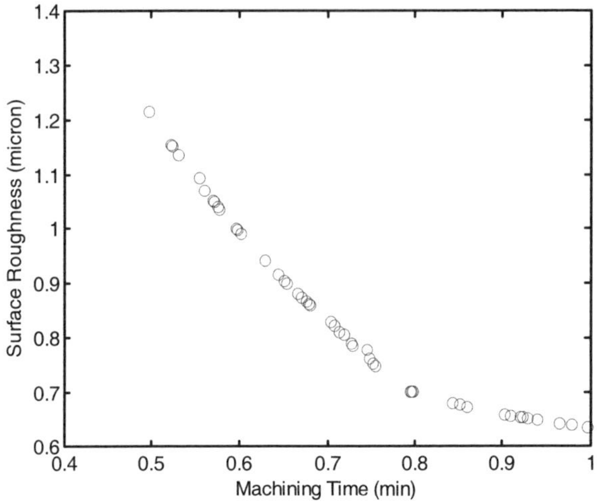

Figure 6.7. *The Pareto front of non-dominated solutions for machining parameters using chamfered inserts*

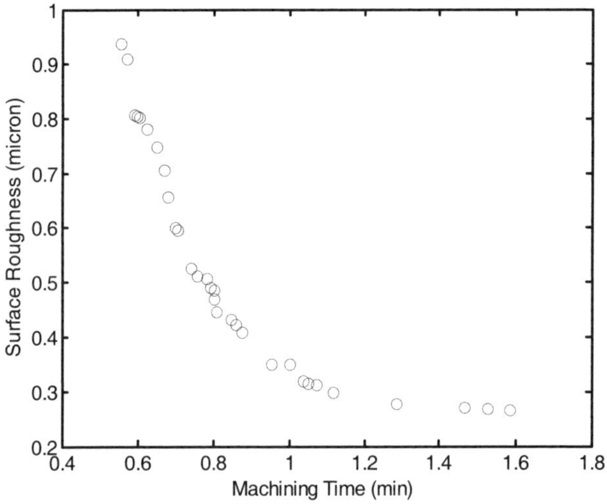

Figure 6.8. *The Pareto front of non-dominated solutions for machining parameters using honed inserts*

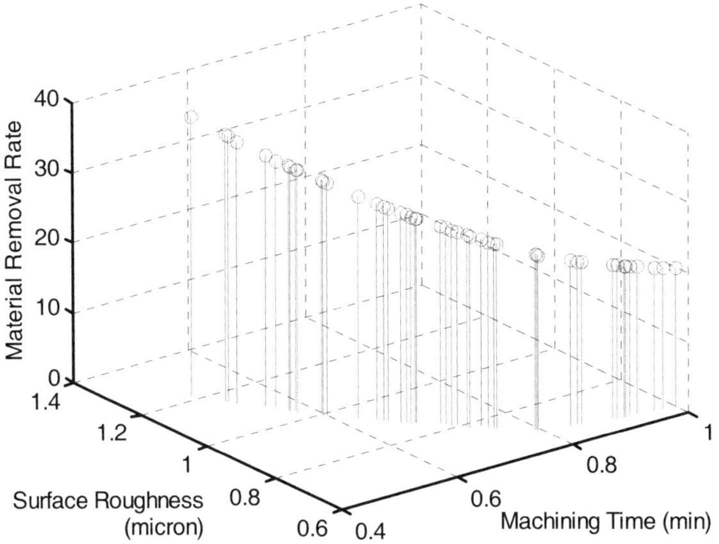

Figure 6.9. *The Pareto front of non-dominated solutions with MRR for machining parameters using chamfered inserts*

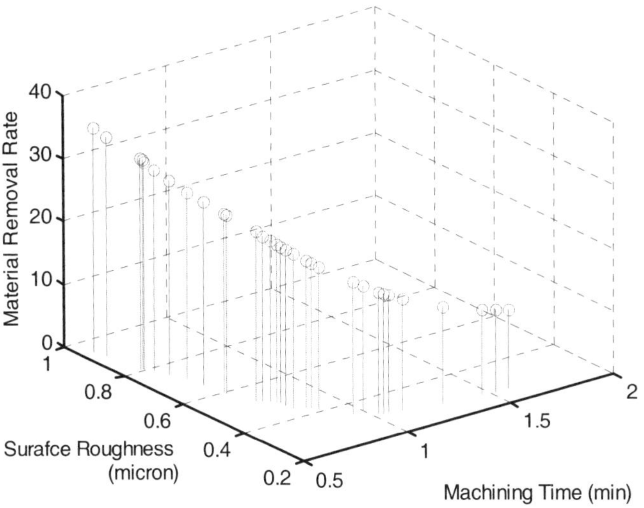

Figure 6.10. *The Pareto front of non-dominated solutions with MRR for machining parameters using a honed insert*

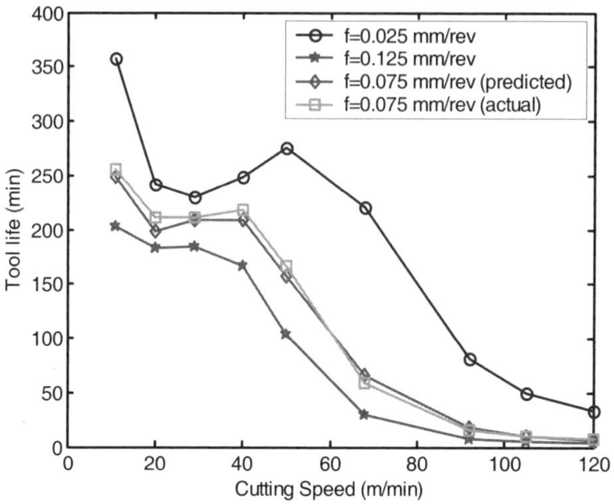

Figure 6.11. *Tool life (depth of cut 0.1 mm) data from [MAM 05]*

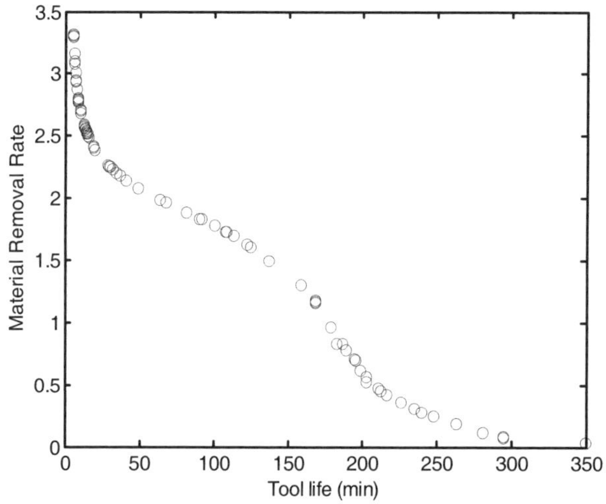

Figure 6.12. *The Pareto front of non-dominated solutions for machining parameters in case study 2*

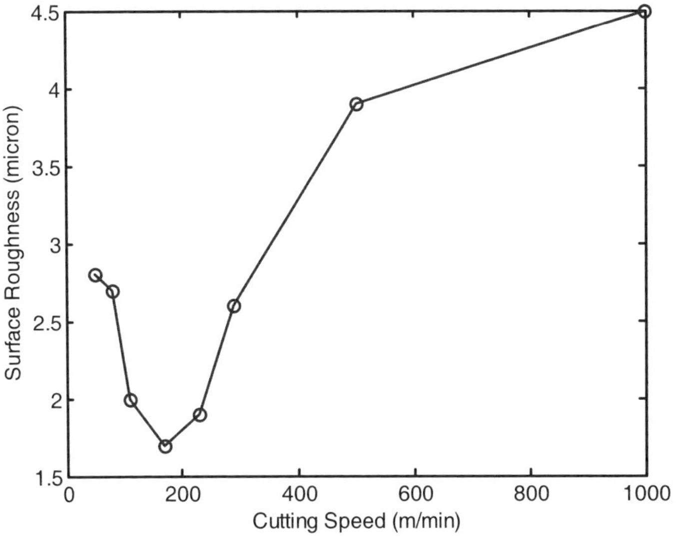

Figure 6.13. *Measured surface roughness of the machined surface for each cutting speed [JAC 02]*

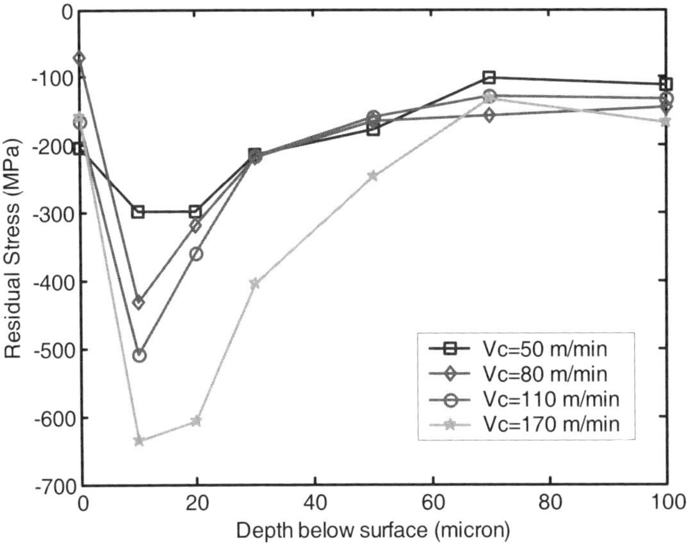

(a)

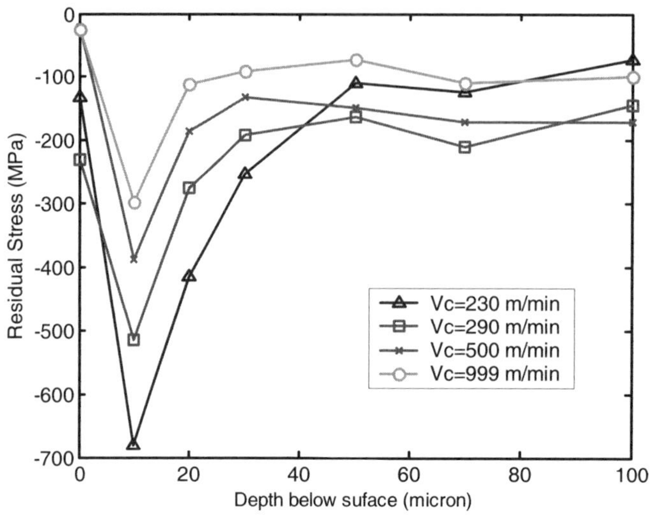

(b)

Figure 6.14. *Measured residual stresses on the surface and the subsurface [JAC 02]*

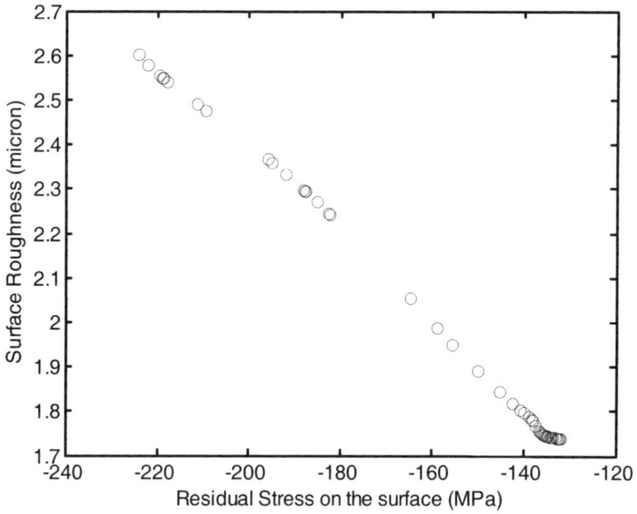

Figure 6.15. *The Pareto front of non-dominated solutions for machining parameters in case study 3*

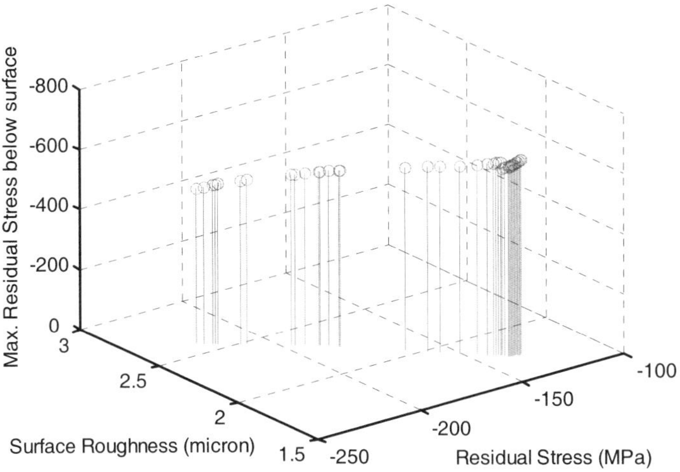

Figure 6.16. *The Pareto front of non-dominated solutions with minimum compressive stresses beneath the machined surface*

6.4. References

[BEN 03] BENARDOS P.G., VOSNIAKOS G.C., 2003, "Predicting surface roughness in machining: a review", *International Journal of Machine Tools and Manufacture*, Vol. 43, pp 833-844.

[CHO 02] CHOU Y.K., EVANS C.J., BARASH M.M., 2002, "Experimental investigation on CBN turning of hardened AISI 52100 steel", *Journal of Materials Processing Technology*, Vol. 124, pp. 274-283.

[COE 02] COELLO C.A., LECHUGA M.S., 2002, "Mopso: A proposal for multiple-objective particle swarm optimization", *Proceedings of IEEE, World Congress on Computational Intelligence*, pp.1051-1056.

[CUS 03] CUS F., BALIC J., 2003, "Optimization of cutting process by GA approach", *Robotics and Computer Integrated Manufacturing*, Vol. 19, pp. 113–121.

[DAH 04] DAHLMAN P., GUNNBERG F., JACOBSON M., 2004, "The influence of rake angle, cutting feed and cutting depth on residual stresses in hard turning", *International Journal of Machine Tools and Manufacture*, Vol. 147, pp. 181-184.

[DIM 97] DIMLA D.E., LISTER P.M., LEIGHTON N.J., 1997, "Neural network solutions to the tool condition monitoring problem in metal cutting – A review critical review of methods", *International Journal of Machine Tools and Manufacture*, Vol. 39, pp. 1219-1241.

[EBE 95] EBERHART R.C., KENNEDY J., 1995, "Particle swarm optimization", *Proceedings of IEEE International Conference on Neural Networks*, Piscataway, NJ, pp. 1942-1948.

[EHM 97] EHMANN K. F., KAPOOR S. G., DEVOR, R. E., LAZOGLU I., 1997, "Machining process modeling: a review", *Journal of Manufacturing Science and Engineering*, 119, pp. 655- 663.

[ELB 96] ELBESTAWI M.A., SRIVASTAVA A.K., EL-WARDANY T.I., 1996, "A model for chip formation during machining of hardened steel", *Annals of the CIRP*, Vol. 45/1, pp. 71-76.

[ELB 97] ELBASTAWI M.A., CHEN L., BECZE C.E., EL-WARDANY T.I., 1997, "High-speed milling of dies and molds in their hardened state", *Annals of the CIRP*, Vol.46/1, pp.57-62.

[ELM 03] EL-MOUNAYRI H., DUGLA Z., DENG H., 2003, "Prediction of surface roughness in end milling using swarm intelligence", *Proceedings of the 2003 IEEE Swarm Intelligence Symposium*, pp. 220-227.

[FAL 99] FALLBÖHMER P., RODRIGUEZ C., ÖZEL T., ALTAN T., 1999, "High speed machining of cast iron and alloy steels for die and mold manufacturing", *Journal of Materials Processing Technology*, Vol. 98, 104-115.

[FIE 02] FIELDSEND J.E., SINGH S., 2002, "A multi-objective algorithm based upon particle swarm optimization, an efficient data structure and turbulence", *Proceedings of the UK Workshop on Computational Intelligence*, pp. 1677-1681.

[HU 02] HU X., EBERHART R., 2002, "Multiobjective optimization using dynamic neighborhood particle swarm optimization", *Proceeedings of IEEE Swarm Intelligence Symposium*, pp.1404-1411.

[HUA 04] HUA J., SHIVPURI R., CHENG X., BEDEKAR V., MATSUMOTO Y., HASHIMOTO F., WATKINS T.R., 2004, "Effect of feed rate, workpiece hardness and cutting edge on subsurface residual stress in the hard turning of bearing steel using chamfer + hone cutting edge geometry", *Materials Science and Engineering*, Vol. 394, pp. 238-248.

[JAC 00] JACOBUS K., DEVOR R.E., KAPOOR S.G., 2000, "Machining-induced residual stress: experimentation and modeling", *ASME Journal of Manufacturing Science and Engineering*, Vol. 122, pp.20-31.

[JAC 02] JACOBSON M., DAHLMAN P., GUNNBERG F., 2002, "Cutting speed influence on surface integrity of hard turned bainite steel", *International Journal of Machine Tools and Manufacture*, Vol. 128, pp. 318-323.

[KAR 05] KARPAT Y., ÖZEL T., 2005, "Hard turning optimization using neural network modeling and swarm intelligence", *Transactions of North American Manufacturing Research Institute*, Vol. XXXIII, pp. 179-186.

[KÖN 84] KÖNIG W.A., KOMANDURI R., TOENSHOFF H.K., ACKESHOTT G., 1984, "Machining of hard metals", *Annals of the CIRP*, Vol.33/2, pp. 417-427.

[LIU 04] LIU M., TAKAGI J., TSUKUDA A., 2004, "Effect of tool nose radius and tool wear on residual stress distribution in hard turning of bearing steel", *Journal of Materials Processsing Technology*, Vol. 150, pp. 234-241.

[MAM 05] MAMALIS A. G., KUNDRAK J., HORVATH M., 2005, "On a novel tool life relation for precision cutting tools", *Journal of Manufacturing Science and Engineering*, Vol. 127, pp 328-332.

[MAR 04] MARLER R.T., ARORA J.S., 2004, "Survey of multi-objective optimization methods for engineering", *Struct. Multidisc. Optim.*, 26, 369-395.

[MAT 99] MATSUMOTO Y., HASHIMOTO F., LAHOTI G., 1999, "Surface integrity generated by precision hard turning", *Annals of the CIRP*, 48/1, 59-62.

[MIT 98] MITTAL S., LIU C.R., 1998, "A method of modeling residual stresses in superfinish hard turning", *Wear*, Vol. 218, pp.21-33.

[MOS 03] MOSTAGHIM S., TEICH J., 2003, "Strategies for finding good local guides in multi-objective particle swarm optimization", *Proceedings of IEEE 2003 Swarm Intelligence Symposium*, pp. 26-33.

[ÖZE 02] ÖZEL T., NADGIR A., 2002, "Prediction of flank wear by using back propagation neural network modeling when cutting hardened H-13 steel with chamfered and honed CBN tools", *International Journal of Machine Tools and Manufacture*, Vol. 42, pp. 287-297.

[ÖZE 03] ÖZEL T., 2003, "Modeling of hard part machining: effect of insert edge preparation for CBN cutting tools", *Journal of Materials Processing Technology*, Vol. 141, pp. 284-293.

[ÖZE 05a] ÖZEL T., HSU T.-K., ZEREN E., 2005, "Effects of cutting edge geometry, workpiece hardness, feed rate and cutting speed on surface roughness and forces in inish turning of hardened AISI H13 steel", *International Journal of Advanced Manufacturing Technology*, Vol. 25, pp. 262-269.

[ÖZE 05b] ÖZEL T., KARPAT Y., 2005, "Predictive modeling of surface roughness and tool wear in hard turning using regression and neural networks", *International Journal of Machine Tools and Manufacture*, Vol. 45, pp. 467-479.

[ÖZE 07] ÖZEL T., ZEREN E., 2007, "Finite element modeling of the influence of edge roundness on the stress and temperature fields induced by high speed machining", *International Journal of Advanced Manufacturing Technology*, Vol. 35, No.3-4, pp. 255-267.

[ÖZE 08] ÖZEL T., KARPAT Y., SRIVASTAVA, A.K., 2008, "Hard turning with variable micro-geometry PcBN tools", *CIRP Annals, -Manufacturing Technology*, 57, 73-76.

[PAR 71] PARETO V., 1906, *Manuale di Economica Politica*, Societa Editrice Libraria. Milan; translated to English by A.S. Schwier as *Manual of Political Economy*, edited by A.S. Schwier and A.N. Page, New York.

[SAR 02] SARAVANAN R., ASOKAN P., SACHIDANANDAM M., 2002, "A multi-objective genetic algorithm (GA) approach for optimization of surface grinding operations", *International Journal of Machine Tools and Manufacture*, 2002, Vol. 42, pp. 1327-1334.

[SAR 06] SARAVANAN R., 2006, *Machining Optimization through Intelligent Techniques*, CRC Press, Taylor & Francis, Boca Raton, Florida, USA.

[SHI 98] SHI Y., EBERHART R.C., 1998, "Parameter selection in particle swarm optimization", *Proceedings of 7th Annual Conference on Evolutionary Programming*, San Diego, USA.

[TAN 02] TANDON V., EL-MOUNAYRI H., KISHAWY H., 2002, "NC end milling optimization using evolutionary computation", *International Journal of Machine Tools and Manufacture*, Vol 42, pp. 595-605.

[THI 00] THIELE J.D., MELKOTE S.N., PEASCOE R.A., WATKINS T.R., 2000, "Effect of cutting-edge geometry and workpiece hardness on surface residual stresses in finish hard turning of AISI 52100 steel", *ASME Journal of Manufacturing Science and Engineering*, Vol. 122, pp.642-649.

[YAN 02] YANG X., LIU C. R., 2002, "A new stress-based model of friction behavior in machining and its significant impact on residual stresses computed by finite element method", *International Journal of Mechanical Sciences*, Vol. 44/4, pp. 703-723.

[YEN 04] YEN Y.C., JAIN A., ALTAN T., 2004, "A finite element analysis of orthogonal machining using different tool edge geometry", *Journal of Materials Processing Technology*, 146: 72-81.

[ZIT 03] ZITZLER E., LAUMANNS M., BLEULER S., 2003, "A tutorial on evolutionary multiobjective optimization", *Workshop on Multiple Objective Metaheuristics (MOMH 2002)*, Springer-Verlag, Berlin, Germany, 2003.

List of Authors

Professor Pedro J. ARRAZOLA
Department of Manufacturing Department
Escuela Politécnica Superior de Mondragón Unibertsitatea / Mondragon University Arrasate/Mondragón
Spain

Professor J. Paulo DAVIM
Department of Mechanical Engineering
University of Aveiro
Portugal

Professor Joaquim DE CIURANA
Department of Mechanical Engineering and Industrial Construction
University of Girona
Spain

Professor Vinayak N. GAITONDE
Department of Industrial and Production Engineering
B.V.B College of Engineering and Technology
Hubli, Karnataka
India

Professor Wit GRZESIK
Faculty of Mechanical Engineering
Department of Manufacturing Engineering & Production Automation
Technical University of Opole
Poland

Professor Ramesh S. KARNIK
Department of Electrical & Electronic Engineering
B.V.B College of Engineering and Technology
Hubli, Karnataka
India

Professor Luis Norberto LOPEZ DE LACALLE
Department of Mechanical Engineering, Faculty of Engineering
University of the Basque Country
Alameda de Urquijo
Bilbao
Spain

Professor Tugrul ÖZEL
Department of Industrial and Systems Engineering
Rutgers University
Piscataway, New Jersey
USA

Professor Ramón QUIZA
Department of Mechanical Engineering
University of Matanzas
Cuba

Index

G, H, I

L, M

N, O, P, Q, R

S

T, U, W